Why
BALLOONS
RISE *and*
APPLES FALL

Why
BALLOONS
RISE *and*
APPLES FALL

The Laws that make the World Work

Jeff Stewart

MICHAEL O'MARA BOOKS LIMITED

First published in Great Britain in 2010 by
Michael O'Mara Books Limited
9 Lion Yard
Tremadoc Road
London SW4 7NQ

A CIP catalogue record for this book is available from the British
Library.

Papers used by Michael O'Mara Books Limited are natural,
recyclable products made from wood grown in sustainable forests.
The manufacturing processes conform to the environmental
regulations of the country of origin.

ISBN: 978-1-84317-413-4

1 2 3 4 5 6 7 8 9 10

www.mombooks.com

Illustrations by David Woodroffe
Designed and typeset by Ed Pickford

Printed and bound in Great Britain by Clays Ltd, St Ives plc

CONTENTS

ACKNOWLEDGEMENTS

This book is dedicated to the amazing Vix.

Thank you Mum and Dad, and Rosemary and Terry,
for helping create the space-time I needed to write.

Thank you Herbie, Iola, Rufus and Oscar for
making sure I took regular breaks.

Thank you Lindsay for making the
whole thing happen.

And thank you, Charles Symons, for your careful
reading and perceptive suggestions.

1

A Brief History of Physics

What exactly *is* physics?

Well, in a way, it's everything.

Physics aims to tell us about a Big Bang that created the universe long ago, to explain how people got here from there (and why we won't be going anywhere else in a hurry), and to show how and why everything around us works as it does.

It tells how the first bits of matter appeared, how the first stars were born, and how, over billions of years, the universe came to be the vast and amazing place we know, with our planet an insignificant speck on the edge of one galaxy in 125 billion.

It explains almost everything that happens in the world around us: energy and movement, sound and light, electricity and matter. And its laws form the basis of chemistry and biology.

Physics also suggests exciting new ideas. For example, it says that time travel may be possible. Unfortunately, it also says that we're probably too big to have a go.

In short, modern physics gives us a fascinating, awe-inspiring and sometimes downright weird view of the universe and our place in it.

Falling apples and rising balloons are just the start.

It's the law

Physicists – a mixture of mathematical thinkers and more practical types who enjoy doing experiments, such as smashing tiny bits of stuff into even tinier bits – think that they can explain all this because everything happens according to laws of nature.

These laws show that, if this happens, then so will that. If I hang a weight (me, for example) from a spring, the amount it stretches will be proportional to my weight: double the weight, double the stretch. (This particular law is known as Hooke's Law because it was discovered by the seventeenth-century British physicist Robert Hooke.)

The laws of physics are useful to us because physics is a practical science. What we've learned has helped us build everything from bathroom scales (that spring again) to a billion shiny gadgets, from the bulbs that light up our cities to the aircraft that don't (usually) crash down on them.

Of course, it has also brought us enough nuclear warheads to blow all of this – and us, and life as we know it – to bits.

How we found physics

People have always tried to explain and predict the world around them. It seems to be an essential part of what makes us human. But it wasn't until we got past blaming everything from lightning to earthquakes on a bunch of irritable gods that we started getting anything useful out of our explanations.

So that's what physics is. But a quick run through two and a half millennia of scientific progress will give us a better idea of how it got us here. And, hopefully, prove to anyone still frightened by the subject that it doesn't bite – even if it does bang.

It's all Greek

Almost 2,500 years ago, the ancient Greeks did a lot of thinking about science. As well as running around naked at the first Olympic games, writing tall tales about gods, cunning heroes and many-headed monsters, and building wonderful temples, they came up with plenty of interesting theories.

For example, Thales supposed that all the earth floated on water, so that earthquakes were caused by waves. Aristotle, whose *Physica* is the first work on physics to use the word in the title, believed everything in the world was made up of earth, air, fire and water, with the heavens made of a divine substance called aether. Smoke, he thought, rose, because it was mainly made up of air, and air always tended to be above earth.

These were nice simple theories, but the Greeks generally argued that an object does something because it's the kind of object that does that kind of thing. This gets us nowhere: it's a circular argument; a good example of what modern physics isn't. (Today, we try to explain things in terms of other things, which is a nice way of saying we usually like to blame someone else.)

Strictly speaking, the Greeks said that, for example, all circles we see are somehow shadows of the 'Form' of roundness. Forms were supposed to be divine, perfect and not of this world, so that the Form of roundness set the perfect example for all other round things. But, it seems, there's no getting round the fact that an object's roundness is still explained by the fact that it is, well, round. Which is, as we noted, circular.

Mind games

Part of the problem was that the ancient Greeks tended to think that you should be able to work out what happens in the world through the power of thought alone. The world around us was imperfect, so, they thought, there wasn't much point looking at it too closely and expecting it to behave in a regular way.

And even when they did observe what was going on around them, they made some strange mistakes. For example, Aristotle made detailed studies of plants and animals, but thought that, as a rule, men have more teeth than women.

Predicting the planets

A few hundred years later, another ancient Greek called Ptolemy (ignore that first 'p' when you say his name or you may be accused of spitting) came up with a fairly accurate mathematical system for predicting the position of the stars and planets.

This was a great step forward for science, except he also thought that the planets (and the sun) revolved around the Earth. This meant that, to make his numbers work, he had to predict that the moon would sometimes come twice as close to the Earth as at other times. As a result, we should, he thought, regularly see the moon appear to double in size.

Of course, the moon never grew, but, in the Western world, Greek ideas held sway in physics for 1,500 years. Partly this was because the church supported them: Greek thought, like Christianity, put man at the centre of the universe, and, with the stars all fixed to the inside of a huge sphere that contained the universe, it also left plenty of room outside for heaven and hell.

As Greek civilization declined, many of its ideas and writings were saved by Islamic scholars, who refined them and slowly increased the importance of maths and observation, particularly in the study of light, the stars and motion. This was a good move because, as it happens, maths and observation turned out to be the key to progress in science.

The scientific revolution

It may seem obvious now but this new way of thinking revolutionized the way we understand the world. People started to discover that by looking carefully, measuring time and distance and energy, and by putting the numbers together with maths and careful thought, we could predict – at least in the laboratory, under carefully controlled conditions – what would happen next and why it would happen. (Actually, it's amazing how far you can get without maths, as this book proves. You need maths to write your theories neatly and to prove them to other scientists, but usually you can understand the big ideas pretty well without it.)

Astronomy also progressed, with the theory that the Earth revolved around the sun slowly gaining favour, despite the efforts of the Catholic Church. The Bible says the world is firmly fixed, and this revolutionary new theory was, indeed, revolutionary.

The great Italian physicist Galileo Galilei, who was handy with a telescope and a dedicated star watcher, offered the first direct evidence for the new theory. For this heresy the church had him placed under house arrest in 1633. He remained there until he died.

Then came Sir Isaac Newton – an English mystic, alchemist and theologian – a complicated and often very cross man who was also the greatest mathematician and second-greatest physicist ever. In one book, the *Principia Mathematica*, published in 1687, he laid down the law of gravity, as well as three elegant laws of motion that describe how objects move. These were unchallenged for two hundred years and still form the basis of most movement calculations by scientists and engineers.

For example, if you want to know the minimum distance in which you can stop your car when you're doing 80 mph on the motorway, and you don't want to cause a pile-up trying it out, all you need is one quick experiment at 20 mph, a copy of Newton's laws of motion and, for the arithmetically challenged, the ability to multiply and divide on a calculator (see page 55).

A clockwork universe

After Newton, it seemed that the universe ran to a hidden clockwork code, and that it was governed by rules that we can discover and understand. This idea changed the course of human history. It led scientists all over the world to use Newton's mathematical tools to build on his ideas and develop science. And so Newton began the rush of scientific discoveries that continues to this day.

We found out why the Earth goes around the sun and how the sun's heat reaches us. We learned about light, sound, electricity and energy: how it changes, how it moves, and how it doesn't disappear. We found that everything we see is made up of tiny bits called atoms, which are made up of tinier bits called electrons and protons and neutrons. And that even then there are tinier bits in them.

With this information to hand, engineers began building the modern world.

At the speed of light

Through the nineteenth century scientists began to discover tiny flaws in Newton's laws – and the branch of physics known as Classical Mechanics, which flowed from them. For example, the orbit of the planet Mercury seemed wrong: its orbit didn't change exactly as the equations said it should. There were also difficulties with handling the speed of light and the way objects radiate heat.

It took the genius of a German patent clerk called Albert Einstein to begin to solve these riddles. In one year, 1905, he wrote four papers that helped physics take huge steps forward – and all while holding down his job at the office. His new ideas included the theory of special relativity (which showed we could never travel faster than the speed of light) and mass–energy equivalence (the famous $E = mc^2$ equation, which showed we could turn objects into energy, and paved the way for nuclear power and atom bombs).

Einstein's theories agreed very closely with Newton's laws under everyday conditions and showed why Mercury went where it did. They also suggested a whole new direction for physics, now called cosmology, which takes it out to work on the entire universe, to tell how stars and even universes might be born.

A hundred years later Einstein's physics still points the way in cutting-edge research.

Neither here nor there

In 1900, a German physicist, Max Planck, solved the radiation of heat problem when he showed that the energy of electromagnetic waves is quantized, so that heat energy is emitted in chunks, which can only get so small and no smaller. His insight led, in the 1920s, to quantum physics, a whole new branch of the science, which was discovered and fleshed out by a large group of physicists, working at the same time, in different countries, on what often seemed separate problems. Together, they made quite a mess of the comforting idea of a predictable clockwork universe.

Quantum physics is the science of the very small: of individual light particles, the structure of atoms and all the bits that make them up, not just the electrons and protons that we already knew, but the tinier bits of these bits, such as quarks, all of which have the odd property of being really just waves that are probably here and probably there, though we can never quite be sure.

New questions

And so we reach the present day. The main thrust of physics today is to put together the theories about the big stuff (Einstein's relativity) and the tiny stuff (quantum physics) into one explanation of the whole universe: how it works, why the laws are as they are, its birth in a Big Bang and, maybe, its death, billions of years in the future.

One version of this goal is called a GUT: a Grand Unified Theory. But whatever it's called, at the moment it's got physics pretty well stumped. The bad news is that it may actually be impossible to create such a theory; or possible, but not possible for human brains.

No one's giving up yet, but these are some of the reasons that some physicists are thinking carefully (or possibly staring out the lab window and daydreaming) about what kind of things their laws are, and how they fit into the universe.

Wrong again

Of course, physics – and its laws – aren't nearly as straightforward as this potted history might suggest.

Apart from risk of nuclear armageddon, the greenhouse effect and the time we waste working out which of the many shiny gadgets it has spawned we should buy next, there is also the fact that – as we've seen throughout its history, and despite all the trust we put in it to stop our cars – sometimes physics is wrong.

For example, Aristotle thought that heavier things fall faster, which seems to be what we all see: think feathers and lead weights. Except that Galileo and Newton proved him wrong.

Then Newton, with the help of falling apples, used the theory of gravity to explain why things fall, and to predict how quickly they gain speed. That was great: but a few hundred years later Einstein made everything much more complicated.

His theories of relativity say that weight and speed of moving are linked, but in a strange way: the faster things move, the heavier they get. You would have to be moving many thousands of times faster than a jet aircraft before you'd really notice the weight you put on, but experiments have shown that the theory is true.

So, Einstein's theory is true – and, on the other hand, sometimes it isn't. In quantum physics, where we look at the motion of tiny and often very light objects (like, for example, bits of light called photons), it doesn't quite work.

It turns out that our current best theories about tiny particles are incompatible with relativity. Somehow, somewhere, we're still a bit wrong.

Solving the problem

Partly, these mistakes are explained away by progress. As we get better at physics, at coming up with clever experiments and measuring the results more and more accurately, our theories slowly get better.

A good theory should always make predictions about things we can see happening (for example, if we double the weight we hang on it, our spring will double in length). If we see a prediction come true, that supports the theory. If we don't, it is disproved, and we have to try again. Every time we see what we expect our confidence in the theory is increased, but it only takes one counterexample to destroy the theory. At least it does in theory.

Actually, when a physicist – let's call him X – gets a 'wrong' result, science has a problem. One way to solve it is to adapt and improve the theory, without changing its basic rules, so that it predicts X's results, too. That way we don't have to throw away whole, useful theories just because that troublemaker X was up to his experiments again.

When it comes to springiness, we say that Hooke's law only holds for certain kinds of material, under certain conditions – to be precise, it holds when springs are springy. Pull your spring too hard and it reaches a point where it turns into a twisted bit of wire. Keep pulling and it'll break. But breaking a spring doesn't break Hooke's law, because the law doesn't apply to bits of bent wire.

But even improving a theory, setting limits and exceptions, is pretty hard work. One easy alternative is to accuse X of getting his results wrong and wonder why he gets so much money for his wrongheaded research into cold fusion and luminiferous aether. There's always plenty to argue about in science – for researchers that's half the fun of it.

Revolting physicists

Sometimes, the little adjustments and strange results begin to mount up, as they did for Newtonian physics in the nineteenth century. Our best theories become a hodgepodge of exceptions and special cases, and arguments rage about unexplained results and experimental error.

At this point it takes a revolution to sort things out, to explain a host of problems, and give us a whole new way of seeing the universe. The arrival of quantum physics was just such a revolution. Newton's universe was solid, precise, and ran like clockwork. Quantum physics says this is just how it looks to us, but that underneath it is flimsy and never completely predictable.

In the same way, some think that the current struggle to produce a unified theory may be a signal that relativity and quantum physics are missing something important, too. No one knows what the next revolution might look like.

Fundamental physics

The hope of physicists is that these revolutions give us a deeper understanding of our world and how it works. Hooke's law is fine, but we now also have complicated theories (about how atoms move in a piece of metal and the law of conservation of energy) that explain Hooke's law, and what happens to a piece of metal when Hooke's law stops applying, how it stretches permanently and eventually breaks.

So there are laws that explain other laws. And somehow the explaining laws (which should be equally testable in experiments) are more important and deeper than the others. At the bottom of all the layers of explanation we think that there may be just a few fundamental laws that will prop up all of physics and the universe we know.

Even now, all our best attempts at the deepest laws, which seem to explain almost everything that happens in the universe and all the other laws we know, could be written, in mathematical symbols, on a single sheet of paper.

One goal of physics is to reduce that sheet to a single mathematical sentence that explains everything. Another is to get it spot on.

Putting it all together

The idea of fundamental laws leads to deep questions about what exactly physics does. Physicists tend to believe that there are real laws to discover, but they find it difficult to answer questions about what they are and how they work.

We'll come back to these debates, and what they may tell us about our place in the universe, at the end of this book, when we've had more of a look at the laws themselves.

For now, we'll note that physics tries to fit together a picture of the whole universe, which means that to understand one bit you need to understand another, but to understand that bit it helps to grasp the first. So don't give up if you don't get one bit. You might understand after reading something else.

Mad (and cross) scientists

It's also worth remembering that physics – like football, fashion, fishing and even things we do that don't start with 'f'– is a human activity. Physicists are people. They argue about their ideas, they make mistakes, they do what needs to be done to make sure they have money to carry on their work next year. Many of them work obsessively at their ideas, and get very excited about them. And it's only with a dose of hindsight, when we forget the dead-ends and the debates, that the story of physics seems neat and nice.

Maybe another way to put it is that many physicists are, quite frequently, a bit nutty. Einstein, for example, liked to go sailing when there wasn't any wind ('for the challenge,' he said). Newton, though, wasn't so much mad as just very cross (possibly because the mercury he worked with got to his brain). He went to extraordinary lengths to get his scientist mates to gang up and belittle some of his rivals.

Nikola Tesla, the gifted Serbian physicist and engineer who made huge advances in the study of electricity, probably had obsessive compulsive disorder. He insisted that the number of his hotel room must be divisible by three; he didn't like to touch round things; and he kept saying his physics theories were right after he moved to the United States and ended up mending roads because no one believed him. The thing is, he was (mostly) right.

Nuttiest of all

But the best mad physicist was an ancient Greek called Archimedes. It seems unlikely that he actually ran naked down the street shouting 'Eureka!' (which means 'I have found it!') because he was so excited after noticing that the water level rose when he stepped into his bath. But then again, in doing that he had just discovered the answer to an important question asked by his king, Hiero of Syracuse, and he was then able to prove that a crown the king had been given was a fake.

Archimedes is important because he reminds us that, in physics, we can all enjoy 'Eureka!' moments.

We can all appreciate the everyday – but amazing – physics of rainbows, and make a good argument for our belief that the world is round though it seems flat. And with a little effort (a bit of peace and quiet and maybe a stiff drink to see us through) we can even think through a thought experiment that demonstrates something as odd as Einstein's theory of relativity.

Read on, and see for yourself. Just remember, at all times, to keep your clothes on and your wits about you.

2
Forces

In this chapter, we'll find out how pushing and pulling change the way things move according to Sir Isaac Newton's three laws of motion; and how sometimes no matter how hard you try you won't get anywhere.

O K. So let's get moving.

Four hundred years ago, as the scientific revolution started putting the whole world in a spin, Galileo, Newton and other early physicists spent much of their time looking at how and why things move. And if forces were a good enough place for Galileo and Newton to start, they should be good enough for us, too.

They're also a good place to start because we intuitively have a feeling for forces and motion. They are part of the physics that we use and control every day: for walking, lifting, carrying, driving and playing with hammers. And, occasionally, forces and motion remind us how big and wild and implacable are the laws of the universe. For example: don't play with hammers, you may drop one on your toe.

So we know that the faster we drive the longer it takes to stop, or that, when it's just you in the car it goes like a rocket, but fill it with luggage and your extended family, and it'll be losing out to a snail in the race from the lights.

Sometimes, though, our intuitions about forces and motion mislead us. Let's look at an example.

Float like a feather, fall like a hammer

We all know that something as light as a feather (like, for example, a feather) will fall more slowly than something heavy, like, for example, a hammer. Except, sometimes it won't.

In July 1971, when Dave Scott, commander of NASA's *Apollo 15* mission, stood in the desolate moonscape and dropped a hammer and a feather at the same time, they fell together, and hit fine powdery moon dust together. And since it cost the American taxpayer around $150 bn (at today's prices) to get him there, we'd better hope we can learn something from that.

Of course, if you drop the feather on Earth, it doesn't fall straight down, but floats in the breeze. Actually, the breeze is the clue to what's happening. On the moon, there's no air to slow the feather. Taking the air out of the equation by doing the experiment in a vacuum on the moon is a great example of the way physics simplifies problems so that we can find out exactly what's happening.

Anyway, our experience with light things like feathers still makes it difficult to get a feel for the physics. Somehow, even after all Scott's effort, it still seems as if a 2-kilo weight will fall faster than a 1-kilo weight.

Well it won't. And here's how to prove it in your head using a thought experiment.

For this thought experiment you'll require: three imaginary apples.

Imagine dropping three apples, all at once, from shoulder height to the floor. (It's easy: just imagine you have three strong arms to hold them. And imagine you're really tall so that they take a good while to fall. See, you can do this physics stuff.)

So, all three apples hit the ground at the same time. Of course, they did. Now what if we loosely tie one apple to another with a handy bit of imaginary string. And drop all three again. Again, all three bang down together. Of course, they do. How could loosely tying two apples together make them fall faster?

Imagine making that bit of string shorter. Does that change the speed at which the tied apples fall? Of course not. Even shorter? Still no change. In fact, even if we get a whole roll of sticky tape and turn two apples into one big messy apple bundle it won't suddenly start falling faster than its pristine comrade, even though that messy bundle is now twice as heavy.

So that's it proven. Heavier things don't fall faster. Enjoy your apples.

And let's head back to the moon.

The gravity of the situation

Have you ever seen pictures of Neil Armstrong or any of the other lunar astronauts out on a moon-walk? Have you noticed their strange gait, how they lope about and pull off huge two-footed kangaroo hops?

Although, on Earth, heavy things and light things fall

at the same speed (if the breeze doesn't get in the way), and the same is true on the moon, there is an important difference between what happens here and up there. It turns out that the actual rate at which things pick up speed when falling (their *acceleration*, to use the proper physics word for getting faster) is less on the moon than it is on Earth. So, even in a vacuum, it isn't how heavy things are that decides how quickly they fall, it's where you are when you do the experiment.

In fact, moon-walking astronauts could jump six times as high as they could on Earth. The reason why has got nothing to do with Michael Jackson's famous dance move and everything to do with a *force* called *gravity*, which is much stronger here than on the moon. Gravity, of course, is what sticks us all to the surface of the Earth and makes everything the right way up, even upside down in Down Under.

But what's a force?

Feel the force

We can't see forces. But we can feel them and see their effects. The force of gravity keeps the Earth travelling around the sun and the moon travelling around the Earth, while other forces inside atoms hold them together and help create the matter that makes up everything around us. Forces make your car go faster and slow it down, push it round corners and stop it sinking into the road.

Pushing, pulling, turning, squeezing and stretching all involve forces; we see objects move differently or change shape because of these forces every day of our lives.

Newton's first law

Isaac Newton summed this up in his first law of motion:

Things stay at rest, or continue moving in a straight line at a steady speed, unless acted upon by a resultant force.

This is a beautifully simple law. But that doesn't mean you don't have to be careful with it.

Direction makes a difference

OK. We're in that car at the start of the chapter. Lugging the family away somewhere nice on holiday. We're doing a steady speed of 30 metres per second (also known as 67.1 miles per hour), on a straight section of the motorway, in the outside lane, probably with a long queue of people in a hurry behind, all cursing and complaining and wishing we'd move over so they can smash the speed limit and get wherever they're going ten minutes earlier. But, hey, this is an important experiment, and 30 m/s is a nice round number, so they can wait.

We're travelling at a steady speed; in a straight line. Are there forces acting on our car?

Try winding down the window and sticking your hand

out. Phew! That's a wind. It nearly blew your hand off. So there's a pretty strong force (wind resistance) slowing the car down. But doesn't Newton's law talk about forces making things speed up or slow down?

True. It does. But there is one vital word we haven't examined yet.

Things stay at rest, or continue moving in a straight line at a steady speed, unless acted upon by a **resultant** *force.*

Wind resistance (along with a few other forces, such as friction in the wheel bearings) is slowing your car. But the engine is working to overcome that resistance and keep it rolling at a steady 30 m/s. And the result of these forces, one pushing forwards, and an equal bunch pushing back, is no overall resultant force acting on your car. They cancel each other out and you keep up a steady speed.

Forces acting on a car

Not only does this tell us how clever Newton was to put the word 'resultant' in, it also points to an important aspect of forces: they have *direction* as well as size, or as physicists like to say, forces are *vector* quantities.

Scalar quantities, like volume, on the other hand, have size only. A litre of milk is purely a litre of milk. Add it to another litre of milk in a big jug and you have two litres of milk; there's nothing else to think about when you do that sum.

Vectors are more complicated. If I'm pushing a car with a force of one Newton (that's right, Sir Isaac was such a genius they named the unit of force after him), and you're pushing with a force of one Newton, neither of us is pushing very hard. And, more importantly, because I haven't told you which way we're pushing, it's not clear whether there is a force of two Newtons acting on the car, or none at all.

If we're both pushing in the same direction you can add our pushing power: there is a two-Newton force on the car, and if we're lucky we might move it. But we could be pushing in opposite directions and the resultant force would be zero. And in between those two extremes we could be pushing more or less in the same direction, or almost in opposite directions, and then we could work out the exact angles of our pushing and ask a passing physicist to calculate the exact size and angle of the resultant force on the car.

Balancing acts

Remember our floating, blown-around feather? Well, we should now be able to understand why, even though on the moon it falls as quickly as a hammer, on Earth it doesn't.

Feathers are, light, flat and wide, so wind and air resistance have a large area to push on. This means that, as a feather falls on Earth, air resistance quickly creates a force that is big enough to balance the small force due to gravity that's pulling the feather down. And once those forces are balanced the feather will float down at a constant, slow speed.

Hammers, on the other hand, have a small surface area relative to their heavy weight. So they can fall much faster before the forces of gravity and air resistance balance out.

On the moon, of course, there's no air resistance. The feather and hammer fall at the same speed because the force of gravity is all that's acting on them.

Terminal velocity

Once gravity and air resistance have balanced out, there is no resultant force on an object. No resultant force means no acceleration (remember Newton's first law!), and so the object will continue falling at a constant speed. This speed is known as *terminal velocity*, in other words the final velocity it reaches.

You can clearly see the effect of air resistance in action when you watch a video of groups of skydivers chasing each other and joining hands as they plummet towards the ground. They are experts at changing their shape, and therefore the air resistance acting on them, so they fall more slowly or quickly. In fact, they can vary their speed from about 55 to 90 metres per second (125 mph to 200 mph).

When forces balance

Floating and flying are also good examples of balancing acts. An aircraft stays up because the lift its wings produce is enough to balance the pull of gravity. Similarly, boats float when the upward push of the water – which is due to an effect called buoyancy – equals the pull of gravity.

Archimedes' principle

Archimedes discovered that:

Any object, wholly or partly immersed in a fluid, is buoyed up by a force equal to the weight of the fluid displaced by the object.

So, if you gently float a small boat in a brimful tub of water, and catch all the water that flows over the edge when it is displaced by the boat, you'll find that the weight of the boat is the same as the weight of the water you caught. Neat, eh?

Buoyancy is the reason we feel so wonderfully light and manoeuvrable when we swim. Even a diver with a heavy tank and a weight belt is so buoyed up by the water that he weighs almost nothing as he cruises around the coral reef.

Buoyancy also lifts hot-air and helium balloons in the air. Heated air and helium gas have lower densities than normal cold air. This means they are lighter than air, so a balloon filled with hot air or helium displaces a weight of air greater than its own weight. According to Archimedes' principle, this creates an upwards force on the balloon.

This makes it clear why hot-air balloons have to be so big. A rough calculation shows us that, because air only weighs 1.2 kg per cubic metre, you need to displace at least 80 cubic metres of it to make sure that the displaced weight is greater than the weight of your average hot-air balloonist (who weighs around 80 kg). And to take up 80 cubic metres of space, a cube of air would need to measure 4.5 m along each side.

Since we displace air, too, buoyancy is also acting on us. In air, we weigh about 0.1 per cent less than we would in a vacuum. That is, the atmosphere gives us a weight reduction of about 80 grams (almost 3 ounces) for the average-sized adult! And we don't have to diet to get it.

What Archimedes told us about density

Buoyancy is also useful for finding fake crowns, as Archimedes is said to have demonstrated soon after his Eureka-shouting running-around-naked antics more than 2,000 years ago. All you need is a little understanding of density, to go with what you know about buoyancy.

The density of an object is its mass divided by its volume. It's a measure of how much something weighs per 1-cubic-metre lump.

Material	Density in kg/m³
Air around us	1.2
Ice	917*
Water	1,000
Lead	11,340
Gold	19,300
Core of the Sun	150,000
Neutron star	1×10^{18}**

Archimedes' task was to find out whether a crown given to his king, Hiero II of Syracuse, was really made from gold. First, he borrowed a chunk of gold that weighed the same as the crown. Then he hung the gold from one side of a set of scales and the crown from the other, so they balanced. And finally, while they were still hanging there, he dipped them into a bath of water.

If the crown, like the chunk, had been gold, the scales would still have balanced, because they would both have displaced the same amount of water. But, because the crown was a fake, it was less dense. That meant it had to be bigger to weigh the same amount as the gold. And because it was bigger, it displaced more

---

* *Which is why ice floats.*

** *That's 1 with 18 zeros after it. Be careful where you drop that lump of neutron star. It's super-heavy. Even a teaspoonful would weigh 10 billion tonnes on Earth if you could get it here; which you can't.*

water and was buoyed up more than the gold. So the crown rose and the gold sank (along with, it would seem, the fortunes of the stingy crown giver).

Crown and gold bar balanced on scale

Once in the bath, the gold bar sinks as it is more dense

Checking that a coin is real gold the Archimedes way is smart because, as you can see from the table opposite, it's more dense even than lead, so it's difficult to fake.

Speed and velocity

Let's get back into our car, still hurtling down the motorway. But now, while we're roaring along at 30 metres per second, a side wind hits us.

This new force isn't balanced by the friction pushing the other way (because you are caught up in the physics of the moment and you've let go of the wheel). So now there is a resultant force on the car and it changes direction. It accelerates across the lanes of the motorway, and a dozen cars have to swerve out the way, their drivers shaking their fists.

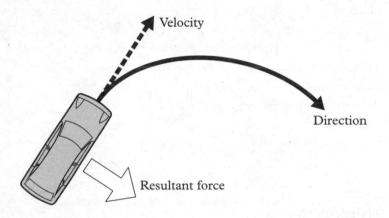

To take account of this, in physics, we talk about velocity rather than speed. Velocity is a vector quantity: like a force, it has a magnitude* (the speed of the car) and a direction (at this point, veering dangerously across the motorway). We tend to think of acceleration as getting faster, but in physics it means any change in velocity, so a change in direction counts as an acceleration.

For example, to keep an object moving in a circle at a constant speed you need a centripetal force: one that is

* Magnitude is just a fancy physics word for 'size'.

continually changing direction so that it is always acting towards the centre of that circle. Does that sound complicated? Because it isn't. Swing a lasso around your head and the rope is just right to provide a force toward the centre of the circle (your hand) and keep the hoop moving round and around (until you let go and the hoop, with no resultant force acting, flies off). In a similar way, gravity, which always acts towards the centre of an object, keeps the planets going round the sun at a constant speed.

Luckily, your car isn't doing U-turns on the motorway, but still your seats apply a force to you and your passengers as it swings across a couple of lanes, so that, instead of continuing straight on at 30 m/s, you all stay inside and turn with it ...

Look out! You're gonna hit that slow coach!

Thinking time

We're all so used to dashing along motorways that we hardly think about how quickly we're really travelling. And motorways are designed for fast driving: there are few features to distract us, bends and hills are long and gentle so we can continue quickly and see a long way ahead, and signs give us lots of warning so we're prepared for what's coming up.

Try doing 30 m/s (nearly 70 mph, remember) in town and you'll soon get a feel for how fast the motorway speed limit is. Or rather, don't. Have a quick look at the table overleaf instead and stay safe.

Speed of light	300,000,000 m/s
Speed of sound	300 m/s
Car on motorway	30 m/s
Person jogging	3 m/s
Child crawling	0.3 m/s
Snail	0.03 m/s

Thinking about your car's speed in unfamiliar units helps you get to grips with it. Every second, your car is travelling 30 metres. Remember, that's 30 long paces. Every second.

When you look at it like this, it's hardly surprising that even a driver who is concentrating can travel 45 metres before slamming on the brakes after they notice that slow coach bumbling along in the inside lane. That's because it can take the human brain one-and-a-half seconds to realize, hey, that's a slow coach, I'm gonna hit it, to then decide to brake and finally, actually hit the brake pedal. And that's an alert driver! If you're thinking deeply about physics you might take even longer.

The upshot? Nothing. We're all in a hurry. But if we were sane we'd be scared witless every time we drive. Especially when you consider that so far we've only done thinking distance ...

Screeching to a halt

Brakes on? Good. But you aren't slowing down very quickly, are you?

There was a reason I told you the car's packed with luggage and your extended family. Didn't you notice when we pulled away from those lights back in town that those snails outpaced us for a bit?

As we started to see earlier, stronger forces cause greater accelerations, for any given object. The relationship is proportional. Double the force, double the acceleration (or, when you apply the brakes, the deceleration). But what if you're using different objects? Well, then you'll find that heavier things accelerate more slowly. Double the mass (mass is a bit like weight – we'll come to it in a moment) and you *halve* the acceleration.

Newton's second law

This is Newton's second law of motion. Physicists usually sum it up by writing:

Force equals mass times acceleration
or
F = ma

(*ma* is nothing to do with your Mum, but is instead a short way of writing *m times a*.)

Plugging some numbers into the equation should make it easy to understand the relationship between these three quantities. If our brakes are trying to stop the car with a force of 7,200 Newtons, and our car has a mass of 1,200 kilos (typical for an empty car), our deceleration would be 6 metres per second every second, because:

$$7,200\ N = 1,200\ kg \times 6\ m/s^2$$

But if we add in four adults and their luggage (an extra 400 kg) we get:

$$7,200\ N = 1,600\ kg \times 4.5\ m/s^2$$

Our deceleration is now only 4.5 metres per second every second. We're slowing much more slowly.

Physicists like to work with numbers and mathematical equations like this (a) because it helps them get useful answers and make their point more quickly and easily than they could with words; and (b) because it frightens off us ordinary folk. But learning equations isn't the same as learning physics. If $F = ma$ is all Greek to you, don't worry. You can still understand forces and how they make things move. You just don't know your a from your α.

Lose weight: climb a mountain

$F = ma$ is also the equation that says we can jump higher on the moon than on Earth, although because it's a special case we replace the a with a g for, you've guessed it … gravity.

So the force of gravity is given by:

$$F = mg$$

Your mass (m) is constant whether you are on the Earth or the moon, because it's a measure of the total amount of stuff that makes you up. Mass is measured in kilos –

which, confusingly, is what most people use to measure weight on Earth. But for physicists, weight is a force. It's the force (F, in the equation) caused by what physicists call the acceleration due to gravity (which is around 9.8 m/s^2 on the surface of the Earth). It's what pulls you down, which squashes the spring in your bathroom scales, which makes the dial spin round to … that much!

Well, never mind. Help is at hand. You could always take a trip to the moon, where g is lower, and so is your weight, which means you get to do all those great big jumps.

Alternatively, you could climb a mountain. Why? Because gravity isn't constant even on Earth. Newton showed that the force of gravity decreases as you move away from the centre of an object: it falls in proportion to that distance away squared. So if you move twice as far from the centre of Earth, the force due to gravity is a quarter of what it was.

On top of Everest you weigh less than you do at sea level – around 220 grams or half a pound less for the average adult! And the walk to the top would certainly do you good: so long as you don't fall off and die, or get altitude sickness and die, or get frostbite and lose your fingers.

Everybody's equal (and opposite)

So, where were we? Safely stopped on the hard shoulder? Good, good.

Now, take a deep breath. Relax. Oh, and tell them in the back to stop that weeping and wailing. Nobody got hurt, did they? They're safe and still now. And luckily, there's Newton's third law to keep them that way.

Newton's third law

This says:

When one object exerts a force on a second object, then the second one will exert an equal and opposite force (of the same kind) on the first one.

You see, forces like to hang about in pairs.

Remember gravity? It's pulling the car towards the centre of the Earth. And the equal and opposite force? Well, it's pulling the Earth toward the centre of the car! Of course, the mass of the Earth is 5×10^{21} (that's 5 with 21 zeros after it) times greater than that of your car, so the effect on the Earth isn't noticeable.

Here's another example of the third law. Remember when we were braking hard as we tried to miss that slow coach? Friction in the brakes applied a force of 7,200 N to slow our car, and so friction on the road provided 7,200 N to stop us skidding. And the equal and opposite force? Well, it's 7,200 N acting in the opposite direction on that bit of the surface of the Earth known as the motorway. So, the other body involved is again the Earth, and although the effect isn't noticeable because the Earth is so massive, our braking does actually affect the way it rotates.

Anyway, enough of these death-defying near-misses. We need to save our energy for something much more important, called – guess what? – energy.

3

Energy and Power

In this chapter we'll talk about all kinds of energy: energy that things have because they're moving, because they're where they are, or even because of what they're like inside. We'll also talk about power, or putting that energy to use. And finally, we'll put it all together and find out how the physics of energy can save us from going out with a bang.

Energy doesn't like to sit around. Which is hardly surprising, given that it's so energetic. Energy is always here, there and everywhere; always doing this and that; always changing into something new. It's like the trendiest party animal in the universe.

As well as propelling us down motorways and into physics experiments, energy keeps us warm, grows our food and helps us see. It even gets us up in the morning and makes the toast.

All our energy comes, initially, from the sun. The sun is one great big long-lasting nuclear explosion, and it's chucking solar energy at us at a fantastic rate. A third of what comes our way is bounced straight back off the

Earth's atmosphere and the clouds and goes back into space, yet our planet still absorbs more energy from the sun in *one hour* than humans burn in *ten months*.

So the sun warms us and the planet (the energy it sends becomes thermal energy). It makes the plants grow, turning solar energy into chemical energy through photosynthesis. We eat some plants, turning their chemical energy into all kinds of other energy, such as kinetic energy (energy you have because you're moving), as we bounce out of bed, and sound energy as we shout at whoever's in the bathroom to get a move on. The sun's energy also evaporates water from the sea and turns it into clouds, and generates the air currents that turn into wind.

Some plants, that were nourished by the sun millions of years ago, have been turned into coal and oil. We dig them up and burn them; they warm us and light our way when the sun's hiding behind clouds or on the other side of the planet. Or we burn them and they shoot us off on holiday, fuelling our airliner as we head somewhere the sun's still shining.

Counting the calories

If you're on a diet, you count calories,* but if you're a physicist you measure energy in Joules.

1 calorie = 4.184 Joules

* The calories you count in your food are *kilogram calories* (often written on food packets as *kilocalories* or, confusingly, simply as *calories*). A kilogram calorie is the amount of energy needed to raise the temperature of 1 kilogram of water by 1°C.

Joules are named after James Joule, a wealthy brewer from Manchester, who was fascinated by electricity. As a child, he experimented by giving his brother and some of his servants electric shocks. But in the 1840s, when he'd grown up, he made himself useful by looking at how he could use the newly invented electric motor instead of steam engines in his brewery. From there, he discovered lots of interesting things about how electrical and mechanical energy is converted into heat.

Eventually it was decided that one Joule would be the amount of energy expended by a force of one Newton if it moved an object one metre. (They try to keep it simple, don't they?)

So, in real life a Joule is the energy it takes to lift a small apple one metre straight up. Or the kinetic energy of a tennis ball lobbed gently over the net. But there are around 720,000 Joules of food energy (carbohydrates and alcohol) in a pint of beer, which is probably why you're going to have to try something more strenuous than the apple-lifting workout if you want to get rid of that beer belly.

Energy conservation

Joule's work on energy conversion led to the discovery of the law of conservation of energy, one of the most basic and important in all physics. Once you know that energy is converted from one kind to another, careful measurement can show that all of it is always converted; that energy is never created or destroyed. Then you have a very important law: the first law of thermodynamics.

It says:

The total amount of energy in a closed system remains constant.

So our universe, the biggest closed system we know, is just as energetic now as it was when it first exploded into existence. And when Mr Joule gives an apple one Joule of potential (height) energy by lifting it up, all that energy becomes kinetic energy as it falls, and then becomes heat and sound (splat!) energy when it hits the floor.

| Potential energy | Kinetic energy |

What a waste

However, converting energy is rarely straightforward. Modern experiments on rowers show that for every five Joules of energy they burn, only around one Joule is actually used to move their oars. The rest goes in heat, in sweating and swearing and making pained expressions for the camera.

And it's not just Olympic gold medal winners who waste energy. Something similar is true of all us when we burn food energy. As a Victorian gentleman, Mr Joule can surely lift an apple without swearing, but he still wastes four Joules of energy doing it.

Most energy conversions are, like Mr Joule's, very inefficient. Even modern low-energy light bulbs only turn half the electrical energy they consume into light. The rest is, again, wasted as heat.

In old-fashioned, conventional light bulbs, only 10 per cent of the energy is turned into light, with the rest warming the room. But even that looks efficient compared to some other energy conversions.

It has been calculated, for example, that for every Joule of energy available in Mr Joule's pint of beer, at least 10,000 Joules of energy have been used in making that beer. That energy is applied mainly in the form of sunlight growing the hops and barley, and in the heating required at various stages as the beer is brewed.

The definition of work

When Mr Joule picked up that apple, he did some work. Not much work, agreed, but certainly a physicist would call what he did work. That's mainly because, in physics, work is defined as 'energy transferred during a process'.

When that apple is lifted it gains one Joule of energy. We know that because it has a weight of one Newton (so it took one Newton of force to lift it) and because we know it moved one metre, and *work done (or energy gained) is force times distance moved in the direction of that force.*

If we drop the apple, that energy is transferred again; gravity does the work to speed it on its way to the ground.

Watts and work

It isn't very exciting that Mr Joule can lift an apple to the not-very-high height of one metre. But what if he can lift that apple quickly? What if every second he lifts an apple a metre, so he's lifting 60 apples a minute, and 3,600 an hour? Well, then he'd be doing one Joule of work per second, which means his power output is one Watt, for no other reason than that is what a Watt is defined to be.

The Watt (W), which is the unit we use to measure power, or how quickly something does work, is named after James Watt, a Scottish inventor whose redesign of the steam engine in the 1700s gave it the power to be the workhorse of the industrial revolution.

Watt a lot of power (all figures in Watts)

Oil-fired power station	output	1,000,000,000
Bryn Titli 22-turbine wind farm on a windy day	output	10,000,000
Lorry	consumption	100,000
Small gas boiler	consumption	10,000
Small one-bar electric fire	consumption	1,000
Bright conventional light bulb	consumption	100
Torch bulb	consumption	10
Mr Joule, apple lifter	output	1

You have more energy upstairs

As you will have realized from all that apple lifting, the higher something is, the more gravitational potential energy it has. That applies to you, too. The higher you go, the greater is your potential energy. In fact, your potential energy is directly proportional to your height off the ground. If you're eight storeys high in a hot-air balloon, you have eight times more energy than when you were only upstairs at home. The amount of potential energy is also dependent on your weight. If you're heavier you don't fall faster, but you will make a bigger splat when you hit the ground.

You can do the thought experiment. We already know

that if you irresponsibly drop from your balloon two apples joined by a loose piece of string, they don't fall faster than one apple. But, compared to one apple, those two do make double the mess and double the bang when their potential energy, after becoming movement (kinetic) energy as they fall, turns into splat energy as they hit the pavement and explode, lightly coating surprised passers-by with apple juice. The law of conservation of energy says that splat energy had to come from somewhere, so it must have come from their potential energy, which became kinetic energy.

So it's clear that an object's gravitational potential energy is directly proportional to its weight (eight apples have eight times as much potential energy as one, which will be released as eight times as much splat energy when they hit the pavement).

It is weight that is important here, not mass. On the moon, in one-sixth Earth gravity, those apples would fall more slowly and splat much less. How much less, we don't, unfortunately, know. Thought experiments only get you so far, and no Apollo astronauts threw apples out the window of the lunar lander.

The law of conservation of energy applies in this experiment because our falling apples are very close to being a closed system. Only a tiny amount of energy is lost to air resistance. At a point when the falling apples are just a few millimetres above the pavement, and just before they splat, they have lost all their potential energy: it has all been turned into kinetic energy. A millisecond later and all that kinetic energy has become splat energy.

Frankly though, finding the kinetic energy of falling apples is not that useful. What about the kinetic energy of a speeding car?

Car crash equations

When you're driving down the motorway and brake hard to avoid a slow coach, your brakes do work (in the physics sense) to slow you down.

Of course, if your brain was working, you'd have remembered all those stopping distances you learned in order to pass your driving test, you would have left a bigger gap to the car in front, and you wouldn't have had to brake so hard in the first place. Let's remind ourselves now.*

20 mph Total = 12 m	
6 m 6 m	➤ Thinking distance
30 mph Total = 23 m	➤ Braking distance
9 m 14 m	
40 mph Total = 36 m	
12 m 24 m	
50 mph Total = 53 m	
15 m 38 m	
60 mph Total = 73 m	
18 m 55 m	
70 mph Total = 96 m	
21 m 75 m	
80 mph Total = 120 m	
24 m 96 m	
90 mph Total = 153 m	
27 m 126 m	

* Yes, I realize it's a bit strange mixing mph with metres, but these are the figures that come from the UK government's transport department (except the 80 and 90 mph lines, as of course those speeds are illegal). It's also notable that these figures quote a much shorter thinking distance than we suggested on page 38. That's because they take into account a 0.7 second reaction time, rather than the 1.5 second thinking-plus-reaction time that modern tests report.

See what happens? If you double your speed, from 20 mph to 40 mph, your stopping distance is quadrupled (from 6 m up to 24 m). Double your speed and you quadruple your stopping distance. Triple it (from 30 mph to 90 mph) and your stopping distance increases *ninefold*. This is obviously one factor that makes speed so dangerous. But why does it happen?

Well, kinetic energy has the answer: but only if we know a bit more about friction, the force that makes your brakes do work. (And it's just as well they do work. Imagine if you ran head-on into a slow coach, or worse, a brick wall. Well the good news is, you don't have to imagine it: we'll get to that, too.)

Get a grip

Friction is a force caused by the resistance of surfaces as they slide over each other. Without it, we'd all be slipping around like non-skaters on an ice rink.

Scientists used to think friction was caused by the bumps and lumps of the surfaces as they knock together. But it seems friction is actually mostly caused by chemical bonding between the moving surfaces: it's more like sticky tape being dragged along a surface, than sandpaper.

Anyway, friction causes motion energy to be transferred into heat and screeching sound energy. Or, as a physicist might say, the heat and sound cause the frictional force (because the transfer of energy away from motion is what causes the force to be felt). Confused? Don't worry. So are scientists. There's still a lot to learn about friction.

But we can all agree that friction is greater the harder you push the rubbing surfaces together, so heavier things feel greater friction: it's easier to slide a single armchair around the living room than a great big sofa full of couch potatoes.

Friction also depends on the kinds of surface involved. It's not just fashion that encourages us to wear rubber-soled trainers rather than leather-soled brogues on the squash court: we wear trainers because they grip better.

Anything that comes between surfaces can affect friction. For example, fluids, such as oil and water, reduce it by keeping the surfaces apart. This is useful inside your car's engine (it stops energy being wasted as your crankshaft turns), but dangerous outside when the road gets wet or oily.

Slip-sliding away

There are two kinds of friction: static and dynamic. Static friction is the force you feel when you're walking carefully on an icy pavement. There's quite a lot of grip there: so long as you move gently your feet don't slip. But as soon as they start to slide, and the much lower force of dynamic friction takes over, you're on your arse. This is one reason why cars have anti-lock (or ABS) brakes. They are designed to release before your wheels slide on the road, and lower, dynamic friction takes over.

Anyway, the work the brakes do, according to the law of conservation of energy, converts all the kinetic energy of the car into heat, through the friction of the brake pads on the brake discs.

For a car braking hard from motorway speeds that's a lot of energy to convert. During hard braking, each brake heats up with the power of a small electric fire, and if you hit the pedal hard a few times in a row, that heat can build and build until your brakes stop working properly (which is why signs at the top of steep hills encourage you to select a low gear and use your engine to slow you).

Now, here's the clever bit. We know that all our car's kinetic energy has been converted into heat as the brakes did work to slow it. We also know, from the 'work' equation we looked at earlier, how much energy that is: it is given by the braking force multiplied by the distance moved. And we know something odd about that braking distance, namely that it quadruples if our car goes twice as fast. Therefore, the kinetic energy of our car must also quadruple if we go twice as fast. So, kinetic energy must be proportional to the speed of our car squared.

The kinetic energy equation

In the 1700s, before there were cars and brakes to slow them down, a Swiss scientist call Willem 's Gravesande performed an experiment in which he dropped brass balls, at various speeds, into lumps of soft clay. He noticed that the depth of the indentation they left was proportional to the square of the speed: a ball dropped at twice the speed of an otherwise identical ball would leave an indentation four times as deep, or if dropped at three times the speed, the indentation would be nine times as deep.

From this came the equation:

kinetic energy = half times the mass of the object multiplied by the speed of the object squared

or, in symbols

$E_k = \frac{1}{2} mv^2$

E_k is kinetic energy (in Joules), m is the object's mass (in kg) and v is its velocity (in m/s). The ½ is a constant to make the units fit.

See how v, the speed (or velocity) of the object, is squared, so that if we double the speed we get four times the energy? And if you're really good at maths you can actually derive this formula from the work equation (*F times d*). That's interesting, isn't it? There's something about our maths, and the way universe works, so that, when we do clever mathematical things to our work equation (*force times distance*) we get our kinetic energy equation. That's the start of how maths and physics go hand in hand.

Even better: now that we have the equation you don't need to look at government figures to work out your stopping times. All you need to do is one experiment, in a deserted car park. Measure how quickly you can stop from 20 mph. Then you can quickly fill in your own table. If you doubled your speed, to 40 mph, your stopping distance would quadruple. If you multiplied it by 3.5, to get 70 mph, your stopping distance should increase by $3.5^2 = 12.25$ times.

Mind that child

Government figures also tell us that, four times out of five, a child hit by a car travelling at 40 mph will die. But if the car that hits a child is travelling at 30 mph, four times out of five the child will survive.

Drivers who don't like speed cameras or speed humps take issue with these figures. How do we know, they ask, the exact speed of a car when it hits a child who has run out into the road? It certainly seems unlikely that anybody has been throwing children in front of cars travelling at various speeds to see how many survive.

Well, even if you don't believe government figures, and skid-mark data, and experiments that were done on dead bodies and animals, and the stories told by the kinds of injuries children receive when tossed into the air by a carelessly driven car, physics might convince you to slow down. Firstly, there are those thinking and stopping-distance figures we've already looked at. At 30 mph, if a child stepped out 25 metres ahead of you, you'd hit them at 19 mph. At 40 mph, you'd hit them twice as fast – at 38 mph.

You can also look at the forces on the child's body. Let's say a child is struck by a car, and carried down the road by it. That means they have been accelerated very quickly to the same speed as the car (because hitting a 20-kg child with a 1,200-kg car won't slow the car down much). But we know from the kinetic energy equation that the force involved in accelerating something is proportional to the square of the speed. So, although our 40-mph car is travelling twice as fast when it hits the child, the force that acts on the child's body is four times greater. And human bodies are very soft compared to cars.

Crash-test hero

Crashing into children is horrible; but crashing into other cars is – apart from being dangerous, noisy and expensive – quite exciting. That's why we like dodgems – and why some people have an unhealthy interest in accidents.

Starting in the 1940s, Dr John Stapp, a US air force officer, carried out lots of research into the forces that human bodies can survive by appointing himself crash-test dummy and driving a rocket-powered sled into what was, basically, a wall. In this way, he proved that the human body can survive accelerations 40 times stronger than gravity – so long as they act for just a short fraction (0.05) of a second.

Stapp played the dummy, but he was sensible enough to wear a seat belt. And, although his crashes broke numerous bones and caused permanent damage to his eyesight (because his retinas became detached), he survived to discover a lot of useful information that helped others design aircraft ejector seats and car seat belts.

Later, his campaigning was a major factor in making seat belts a compulsory feature on US cars. (He also came up with Stapp's law: 'The universal aptitude for ineptitude makes any human accomplishment an incredible miracle.' It explains why so many people didn't and still don't bother using those belts he worked so hard to put in their cars.)

Smashing fun

So we've looked at stopping times and seat belts. But what else can physics tell us about how to make cars safer for humans hellbent on crashing them in thought experiments?

Let's say we build a really strong little car. You can smash it, head-on, into a brick wall, and it'll hardly get a dent.

However, if we think of the work equation and kinetic energy again, it's clear that, when you hit that wall, the fact that your car stops very quickly means that the force required to stop it – and you – is very large.

Let's do the maths.

From the kinetic energy equation, an 80-kg imaginary person called Bob, who has for some unknown reason volunteered to be in our little car, is driving it at 60 mph (27 m/s). As a result, he has:

½ mv² = ½ x 80 x 27² = 29,000 Joules of kinetic energy.

Now, let's say, as Bob drives head-on into a wall, that his seat belt stretches enough to let him move 10 cm forward (that's 0.1 m) as it stops him from crushing his skull on that hard little steering wheel. Then the force required to take away all his kinetic energy is given by:

KE (29,000 Joules, in this case) = F x d (force times distance)
= F x 0.1
So F is 290,000 N.

For an 80-kg man, this is an acceleration 360 times bigger than gravity.

Well, that's no good. Bob's little car might still work, and his body might still be strapped in, but after experiencing a force like that, his brain has been squished into soup as it was stopped by the inside of his skull. Stapp was pretty much OK after acceleration 40 times stronger than gravity; a racing driver called David Purley just about lived after he crashed and experienced 180 g. But at 360 g (in other words, 360 times the force of gravity) you've got no chance.

To survive, you need to take longer to stop. So what modern car engineering has done is to slow things down by adding, to the front of our car, a 1-metre-long crumple zone: an area that is strong enough to take some of the sting out of an accident by crumpling. This converts some of the energy of the collision into heat and sound, and makes it take longer for the car to slow, so that the force on you is reduced.

Add this to our seat-belt stretch and you now have ten times as long to stop as Bob did (up from 0.1 m to 1 m, because the crumple zone shrinks from 100 cm to 10 cm in the accident).

Now, when the car stops, the ten times the distance to stop means the force on you is reduced to a tenth; which means the acceleration is a tenth, which means it is only 36 times the force of gravity. And Colonel Stapp proved you could survive that.

Hooray. The numbers work. So, in you go, but do drive carefully.

Oh, and do belt up. Steering wheels and dashboards are very efficient stoppers. They'll bring you to a halt in about 1 cm. That's ten times quicker than a seat belt: enough to make even a 30-mph crash fatal. And enough to make it unlikely you'd ever get to grips with the momentous stuff we've got coming up next.

4

Momentum

In this chapter we'll find out about momentum and inertia, and how physics shows us that, once something starts moving, that movement never stops. Then we'll go round in circles, and discover how spinning tops and gyroscopes rely on similar principles. To finish up, we'll talk a bit about torque, the force that makes cars with big engines such fun.

In the martial arts movie *Crouching Tiger, Hidden Dragon*, the heroes can break the laws of physics, which makes for all kinds of spectacular floating and flying as they fight. At one point, the rebellious young heroine, Jen Yu, is attacked as she's having lunch. A young tough guy runs at her, but, without standing or bracing herself, she blocks him with her left hand, and he bounces off her. All the while, she holds her cup of tea in her right hand and doesn't spill a drop.

Her move, though handy, is obviously impossible. But why? Her attacker has just as much kinetic energy after she blocks him as before. He rebounds, so energy is conserved, just as it should be. Yet some law of physics must have been broken – otherwise a smart flyweight boxer would by now have taken a few classes in Jen Yu self-defence and become heavyweight champion of the world.

It's clear, then, that the laws of energy alone cannot show why Jen's heroics are impossible. For a start, energy lacks a sense of direction – it is a scalar quantity, like mass and volume. It has size and nothing else, so it can't explain why it's odd that Jen wasn't knocked backwards (and into next week).

But lack of direction isn't energy's only failing when it tries to describe motion. To discover more, let's try a thought experiment on a glacier: let's try to stop it.

Hold the ice

A glacier is a massive, slow-moving river of ice, so it won't mind us giving it a shove back the way it came. Come on, push as hard as you can. Any luck? How about if you bring in a fleet of tow trucks? Well, it's still hopeless. Even hills and rocks can't stop a glacier: it ploughs right on through whatever is in its way, slowly carving out massive U-shaped valleys as it goes. The forces involved are tremendous.

But what if we string a giant waterwheel-kind-of-thing over a glacier, and try to convert its kinetic energy into electricity, as we could with a river. Can we solve the energy crisis using glaciers?

Well, let's continue with our thought experiment and find out. Let's say our glacier is a 1-cubic-kilometre block of ice. Its mass is roughly 10^{12} kg (1 followed by 12 zeros) because water weighs 1,000 kg per cubic metre (or 1 kg per litre). And let's say it moves, like the Byrd Glacier in Antarctica, which is the biggest in the world, at a speed of around 0.00001 m/s. Then its kinetic energy is a not so whopping 50 Joules. Convert that all to electricity and you get a low-energy bulb lit for a few seconds.

Again, energy isn't giving us the full picture. Glaciers are unstoppable – but they have about as much energy as a student physicist after a night's overindulgence on Mr Joule's brew.

What we need is something that describes unstoppable glaciers and lunging baddies.

And that something is called *momentum*.

Don't stop now

Momentum is something we all intuitively understand: it's why you can't stop a car by giving it a shove; it's what makes pool and snooker games worth playing; and, as you'll have guessed by now, it comes up in a law of the universe that makes Jen Yu self-defence impossible.

The idea of there being something in a moving object that keeps it moving has been around since Roman times. But it wasn't until Newton came along that it was properly, mathematically, useful. This is hardly surprising, as momentum is intimately tied up with Newton's laws of motion, especially the third law, as we'll see when we discover that flaw in Jen Yu self-defence.

The momentum equation

Momentum (which physicists represent with the letter P) is calculated by multiplying mass with velocity, and so is given by the formula:

$$P = mv$$

So our glacier from earlier can have huge momentum even though it is moving slowly. (Its momentum, 10,000,000 kg m/s, is big: about the same as that of a giant car transporter carrying 300 cars quickly down the motorway.)

Momentum is why oil tankers take 20 miles to stop. It's a measure of the 'quantity of motion', so it is proportional both to the mass of an object and its speed.

A load of balls

As physicists, we like to simplify things to find out what's really happening. So instead of trying to find a Jen Yu martial artist and a cooperative thug, we can recreate the scene with pool balls. But if we're going to do experiments on a pool table, we want to carry them out on a physicist's *ideal* pool table: one that's perfectly smooth, not bumpy or beer-stained. Then again, ideal pool tables are in short supply. So you may as well find a cosy pub, get yourself a drink, and settle down for some physics.

First up – to play the part of a relaxed and lunching Jen Yu – choose a ball that's blue. Then, send in from stage left – playing the thug – a gently rolling red. (We won't use the white ball because, on most pool tables, it's a different size and weight to the others, and that makes everything less ideal.)

Now, red hits blue, full ball, dead in the middle. What happens? Well, unless you muck things up by putting spin on the shot, the red stops and blue carries on – and it carries on at exactly the same speed, and in the same direction, left to right, that red was moving.

How momentum works on pool balls

Try it again. And again. Try it in a different pub. And another. The result will always be the same (so long as you get the collision dead in the middle of the blue ball).

Now that makes sense. Again, energy is conserved, but now, our new quantity, momentum, is, too. All the energy and all the momentum on our pool table has been transferred from red to blue.

Newton's third law – the one about equal and opposite forces – has been obeyed, too. As the balls collided, the amount that red pushed blue equalled the amount that blue pushed red, and that amount was enough to stop red and send blue on its way.

The amazing thing is that what you have discovered about momentum, right there in front of you on the pub pool table, is one of the most basic laws of the universe.

Exactly right

Energy and momentum are so important that they are mentioned in a bunch of laws that physicists are pretty much absolutely, completely, honest-we-haven't-got-these-ones-wrong sure about. As far as we know, these laws have never been broken. They applied in the beginning, right after the big bang. And they will apply for ever, until the end of time.

They are called the exact laws.

The exact laws are all conservation laws. They say that the total amounts of certain properties in the universe are always conserved, so that these properties can be neither created nor destroyed. (More usefully for those of us who aren't gods, they also apply to small closed systems like our pool table, if we're careful.)

We've already seen the law of conservation of energy. Now you've done the experiment with the pool balls, you have discovered the momentum version for yourself. It says:

The total momentum in a closed system remains constant. It cannot be created or destroyed.

Energy and momentum are always conserved in our universe simply because of its underlying shape, which is the same whichever way you go, and because time ticks on constantly, without changing the laws of the universe. (The other exact laws describe the conservation of slightly more complex properties, such as electric charge, and much more complex properties, such as colour charge (nothing to do with the price of red paint), weak isospin (nothing to do with a gentle wool wash) and – luckily for gamblers – probability.)

So conservation of momentum is there for racing drivers who want to slingshot out of corners, and for rugby players who need to flatten their opponents. It's there for pool players and for people who like to pick holes in martial arts movie fight-scenes.

What's more, and like we saw in the last chapter with energy, momentum can be simply and beautifully useful in everyday physics, because it is easier to use the fact that momentum is never created or destroyed than it is to get involved with Newton's laws of motion, and all those tricky forces and accelerations.

Using your momentum

As you noticed when you were diligently rolling that thuggish red ball into poor Jen Yu blue, it was difficult to get the collision quite straight, so that red stopped and blue escaped.

Often, red, coming in left to right, would kiss blue at an angle – and our nice little experiment would be ruined. Instead of red stopping, it would carry on one way, maybe slightly up the table as well as left to right, with blue going off at a different angle, slightly down, as well as left to right.

When you think about it carefully, what's happening is quite simple.* After all, even average pool players know this behaviour well and it is all, of course, another result of the conservation of momentum. Let's look at this more closely – but first, time for a bet.

If you can gently mark on the table a straight left–right line that red is rolling along, I bet you that, whatever distance from that line (measured at a right angle from it) blue finishes, red will finish the same distance away, but on the opposite side.

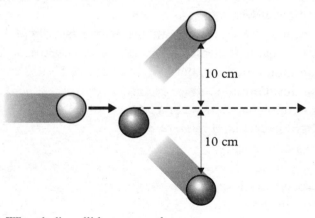

When balls collide at an angle

---

* Good pool players, however, make the physics complicated. They also add in spin: partly to show how brilliant the human brain is at physics that is much too difficult for this book, and partly because that way they can get the cue ball into positions for their next shot that straight momentum would never allow.

Tried it? Was I right? Of course I was. And you owe me a tenner.

Anyway. Back to the physics. So, although both balls go off at different angles, and they travel different distances left to right, at different speeds, if you tell me that red finishes 10 cm above the line, then I know that blue finishes 10 cm below, and all without looking or measuring.

Why? Well (you've guessed it), because momentum is conserved – and, more subtly, because, as we've just seen, momentum has direction, and so it is conserved *in all directions*.

When just red was rolling, it had left-to-right momentum, and nothing up or down. This means that, after the collision, our pool ball system, taken as a whole, still only had left-to-right momentum, because momentum can't be created or destroyed. And therefore, after the kiss, the up momentum of blue must have been cancelled out by the down momentum of red, while the overall left–right momentum that was introduced by red was later shared out between red and blue.

This momentum then gave the balls equal speeds *in the up-down direction* (because they have equal masses), and that speed was lost by both (to friction) as the same rate, and so, in the up-down direction they both travelled the same distance. Good, eh?

'But hold on a moment,' I hear you say. 'Those balls were rolling merrily along and they stopped. And something that's not moving has no momentum. So how on Earth is momentum conserved when it has quite obviously been all used up?'

You can't lose momentum

The clue, in fact, is in the 'how on Earth'.

In the last chapter, we learned – through those risky driving experiments – that, as objects slow and stop, energy is being conserved but, through friction, kinetic energy is converted into heat energy (which, as we'll see later, is in some ways a kind of kinetic energy because it is created by the vibration of atoms). So playing pool actually raises the temperature of that pool table.

Momentum, too, is always conserved. No matter how things bang into each other, or rub up against each other or are magnetically pulled together, the overall momentum of the system is unchanged. But momentum can't change into anything else, like energy can: it's always there, even if it is transferred to something so big (like the planet Earth, for example) that we don't notice its momentum change.

The key to the difference between energy and momentum is once again that momentum is a vector: it has direction as well as magnitude. The vibration of atoms doesn't give an object momentum because those atoms are vibrating randomly in different directions.

Remember how, at the end of chapter 2, we noted that the car's braking affected the spinning of the world, because of Newton's third law? Well, the same law applies as our balls stop. The balls are stopped from moving by the friction from the baize, and the equal and opposite force from the balls pushes the table, and hence the floor, and the country and eventually the whole world!

This is great news for hippies, who get some justification for their belief that everything is connected, man. But it's bad news for them, too, because the effect of a pool ball on the movement of the Earth is so tiny that it is impossible to measure.

In short, hippies can't teach us much about physics. But perhaps Clint Eastwood can.

The good, the bad and the impossible

Just in case you're ever facing a six-gun-wielding outlaw in a graveyard where one of the coffins is packed with cash, it's worth looking at *The Good, the Bad and the Ugly* to find out exactly what won't happen.

First up, it is almost impossible to quickly draw a 150-year-old handgun and hit a person standing 50 m away with a single shot. Heck, with those old guns you'd be pretty lucky to hit a barn door at that distance. But then, even if you got million-to-one-chance lucky, and did hit the baddie, it is then completely impossible for the bullet you fired to knock that baddie over backwards even though the recoil of your gun, which you're shooting from hip height, affects you not one bit as you stand there, cool as Clint.

After all, momentum has to come from somewhere. When your gun fires, and the bullet is blasted down the muzzle, its kinetic energy has come from a small explosion inside the gun. But momentum, as we've seen, can only come from other objects. So what gives your bullet momentum?

Initially, Clint and the bullet in his gun had no momentum. So: after he fired, they can't have had any overall momentum, either. So: whatever momentum the bullet had straight after it was fired, Clint, the firer, should have had the same amount, but in the opposite direction. So: since the bullet hit Lee hard enough to knock him over backwards, the recoil should have knocked Clint back, too, even though he was braced for the bang.

It's been around and it'll keep going around

So far, we've been looking only at one kind of momentum, which physicists call *linear momentum* because it's all about things moving in straight lines. But now it's time for a quick spin around the other kind of momentum: *angular momentum*. It had a large say in the creation of this planet and its rotation. And, because it keeps gyroscopes standing upright, it is also a vital ingredient in the artificial horizon instruments that tell airline pilots when they're flying straight and level – and when they are diving at full speed into the ground.

Like its linear cousin, angular momentum is never created and destroyed. Once there is spin in a system, it's there for ever. In this way, angular momentum is a bit like politics: the spin can be transferred from one object to another and it may be not be noticeable for a while, but it never goes away.

Angular momentum is also important in those slightly dodgy Olympic events, such as gymnastics, diving and prancing on ice, where you get medals for looking nice. Platform divers, for example, have a deep-seated understanding of the fact that, as they spread themselves out, they spin more slowly, whereas as they tuck themselves in, they spin more quickly.

Angular momentum

The angular momentum of an object is the product of its *moment of inertia* (the object's resistance to changes in the way it's spinning) and its angular velocity (the speed at which it's spinning). The equation is written as:

$$L = I \times \omega$$

where L is angular momentum, I is moment of inertia, and ω is angular velocity. (ω is the Greek letter omega. And you thought the linear momentum equation used weird letters.)

Moment of inertia increases as whatever is spinning spreads out or moves away from the axis of rotation. But if that happens, it will spin more slowly, so that its angular momentum is conserved. That's why figure skaters spin slower as they push their limbs out, and it's why, if you're a platform diver, you can tuck yourself into a ball so you spin quickly in the air, and straighten out so you spin much more slowly as you hit the water.

Moment of inertia also increases if the object's mass increases, but even the heartiest snackers in the world would find it impossible to put on weight in the time it takes to fall from a 10-metre-high board.

Defying gravity

Let's try another experiment. First, take a wheel off your bike. Now, hold the wheel with two hands (one on each end of the axle), get it spinning quickly while upright and then let go with one hand.

Come on, trust the physics: let go!

Amazing isn't it? Instead of falling over, your wheel is defying gravity, its angular momentum keeping it upright as you support it one-handed, on one side of the axle. Now you know why cyclists wobble like idiots as they start pedalling away from the lights but are far better balanced once their wheels are spinning quickly: it's all because the spin of their wheels keeps them upright.

Actually, your bike wheel's behaviour is even weirder than it seems at first. After a moment, you'll realize that, instead of falling over sideways, it wants to rotate about the point where you are supporting it. This behaviour is even clearer if you support the wheel on one side of the axle by hanging it from a piece of string. Then, when you get it spinning, it will hang there, defying gravity, spinning, but also rotating gently about the string.

The reason? Well, instead of flat-out defying gravity, the wheel's spin subverts it.

A funny turn

Consider a point at the top of your bike wheel. If the wheel isn't spinning, the falling-over turning force that is applied by gravity and the string makes that top point fall down in an arc to one side. (Physicists, engineers and others like to call turning force 'torque'. More about that in a moment.)

If the wheel is spinning so the top of it is moving away from you, there are no new forces involved, but the spin is already carrying our point at the top of the wheel downwards, except that it is downwards and away from you, rather than downwards to the left.

Yet because the forces on our top point are the same as they were when the wheel wasn't spinning, the laws of physics say it has to get to the same place it would have got if it wasn't spinning. And this is what happens: the spin of the wheel, added to the strange rotation about the string, gets our point on the wheel exactly where it should be, without the wheel falling – and so it doesn't fall.

Effectively, the wheel's spin turns the turning force caused by gravity through a right angle, so that it forces the wheel to rotate rather than fall. Physicists call this phenomenon *precession*: it's not a particularly useful term to remember, unless you need an excuse for spelling procession wrong.

Gyroscopes

The weird spinning-bicycle-wheel-on-a-string contraption you've just been playing with is a crude gyroscope. The great thing about gyroscopes is that the spinning wheel inside them means they're not affected if you try to rotate them. So a gyroscope will balance on the point of a needle, or, more usefully, remain level as you fly your aircraft up and down, and bank left and right.

Combine a few gyroscopes with some acceleration sensors (that measure precession) and you can build a navigation system that helps keep careful track of your accelerations and rotations. And, if you know where you started, that information will always be enough to tell you where you are and how fast you're going. Systems like this are used in aircraft automatic pilots and, more destructively, in long-range missile guidance systems.

Torque isn't cheap

You may have noticed that motoring journalists love the sound of their own voices. But as well as talk, they also love torque.

Torque, as we mentioned, is turning force. It's what causes changes in angular momentum, making an object spin more quickly or more slowly. It's also a measurement of a car engine's ability to turn the drive shaft. So, if the supercar you're testing has an engine that produces a lot of torque the car will be fast and powerful. But the relationship between power and torque isn't straightforward.

Car engines produce different amounts of power as they run at different speeds. They are generally at their most powerful when they are running quite quickly; maybe at around 6,000 revolutions per minute for a petrol engine.

But what motoring journalists love is low-end grunt. If they haven't been eating beans, grunt comes from high torque at low revs: it's the turning force that the supercar's thirsty V8 engine produces as it spins quite slowly; it's what starts the wheels turning; it's what accelerates all that shining metalwork and those swooping lines rapidly and exhilaratingly away from a standing start.

Lots of power at high revs helps you scream down the motorway at more than double the national speed limit; but torque at low revs gets you from standstill to speed limit in 4.5 seconds, your heart pounding as you're pushed back into your seat by the acceleration, a mighty grin on your face as you conquer time and space and creeping middle age.

A sledgehammer to crack a nut

You increase torque by using longer levers to apply the force, or (no surprise here) by increasing the force itself. Physics textbooks like to give nutcrackers as an example of how levers allow us to increase the force we apply to a nut we'd like to crack. You won't crack an almond with your fingers alone, they point out. A nutcracker makes it possible, as long as you squeeze hard.

But we say: if you have a tough nut to crack, try a sledgehammer. Of course, you don't need a sledgehammer to crack even the most obdurate almond, but with one you will get that nut cracked super-quick. The massive weight of a sledgehammer's head gives you lots of force (double the force, double the torque). Now swing it over your head and get that long handle multiplying your force. (You also double the torque if you double the length of the lever.) And now: splat that nut!

Warning: your almond may be too squished to be edible after this experiment, but, sometimes, practicality is not everything – just ask your average supercar buyer.

Yu again

Before we're done with momentum, it is worth admitting that – despite what we found with our experiments in the pub – there *is* a way Jen Yu could have repulsed her attacker, much as she did in *Crouching Tiger, Hidden Dragon*, without breaking the laws of physics.

As Newton's first law showed us, the universe is lazy.

Objects like to carry on exactly as they are unless they get a proper kick up the backside from a resultant force.

To go back to our pool table for a moment: if we roll a ping pong ball into blue Jen Yu, she'll hardly move, but the ping pong ball will bounce straight off her. And don't think that this is because of friction. If we'd taken our blue ball up into orbit, set her floating in the middle of the room, and gently thrown our ping pong thug at her, the result would be exactly the same.

This little experiment shows us two things: first, don't use a ping pong ball as the cue ball if you want a decent game of pool; second, that it isn't only momentum that is important.

The lazy universe

We can say more about this tendency of the universe to stay as it is. We know that the bigger and heavier an object is, the more difficult it is to get it to change its movement (or start it moving if it isn't moving in the first place). An object that isn't moving doesn't have any speed/velocity, so it doesn't have any momentum. But it still likes to remain at rest. And the heavier it is, the lazier it is.

Physicists call this resistance to change in motion 'inertia', which can be a little confusing for the rest of us who tend to equate the inertia of a moving body with its momentum. But it is worth separating momentum and inertia.

Heavy things have great inertia: no matter whether they are moving quickly or slowly, or not moving at all, it is difficult to alter their motion (go push an oil tanker, if you don't believe me).

And light things can still acquire great momentum if they are moving very quickly. The Man With No Name's bullet (if he were a real nameless man, in the real world, firing real bullets, etc) would weigh only 16 grams (a tenth of an ounce). Throwing that at a baddie wouldn't have much effect, but firing it at him at around 300m/s would give him quite a jolt.

It turns out that we measure inertia in kilograms. An object's mass (which, on Earth, because of gravity, gives it weight) also gives it inertia. And for many things that are impossible to weigh (you can't, for example, put a planet on your bathroom scales), a calculation of their momentum, and therefore their inertia, gives us their mass – the measure of just how much stuff they have in them.

We'll see more about the stuff that makes mass in just a moment. After we've tied up one last loose end …

Secret weight problem

So was that *Crouching Tiger* scene realistic, after all?

Jen Yu may have looked like a svelte young kung fu fighter, but did she in fact have a weight problem? We assumed in our earlier pool table experiments that Jen's attacker weighed about the same as her. But if, instead, she weighed much, much more than him, then our new experiment, with the thug played by a ping pong ball, would be a better indicator of the truth of that scene.

So, maybe Jen had a massive weight problem. The kind of massive weight problem where she was something like a couple of tonnes overweight. (The tonne is the metric version of the imperial ton. It weighs about the same but is spelled more creatively.) Then even a heavyweight thug might bounce off her: if she was made out of something bouncy; and so long as her massive weight problem hadn't already caused her to crash through the floor of that rickety restaurant where she was sitting upstairs eating before she (and her attackers) kicked it all to pieces during their fight.

But how did she get to weigh that much eating low-fat rice and noodle dishes and drinking green tea? And how, if she weighed a couple of tonnes, did she manage to leap about like a flea?

Oh well. My advice is, don't watch an action movie if you want to learn physics.

5
Heat and Matter

In this chapter we'll learn all about matter and what it's made of. And then we'll see how what-it's-made-of explains how matter can turn from a hard, springy solid into an airy-fairy gas; why water is weird; why smoke gets in your eyes; and how life on planet Earth keeps us all under pressure.

Matter is the stuff all around us. It takes up space and it has mass: that's why it's filling up your cupboards and making your shelves creak under its weight (though of course those cupboards and shelves, and you and your house, and all the world are made up of matter, too).

Apart from having mass and volume, and creating storage headaches, what else does matter do? Maybe if we find some answers to that, we can find out a bit more about what matter is.

What's the matter?

Thousands of years ago, ancient Greek and Indian philosophers came up with the idea that, if you kept dividing matter up into smaller and smaller pieces, you'd eventually end up with something fundamental: something that just couldn't be split again. The Greeks called that something the *atom*.

But it was a long time before the new science of chemistry began to give these ideas some experimental basis. In 1803, an English scientist called John Dalton noticed that certain elements could be combined with others in a ratio of small whole numbers. So if you had a big, well-equipped chemistry lab and piped 1,000 litres of hydrogen and half as much oxygen into a container, then chucked in a lighted match, there would be a deafening explosion and all the gas would become, once cooled, enough fresh water to fill a large glass.

(If that bang has made the water hard to find amid the general wreckage of your lab, get a mop – we're going to need that glass of water for the next bit.)

If you use alternative volumes of oxygen and hydrogen (for example, the same volume of each, or twice as much oxygen as hydrogen) there will always be some hydrogen or oxygen left over. That's because water is H_2O: water molecules – the smallest bits of water – are made up of two hydrogen atoms and one oxygen atom. So you need twice as much hydrogen as oxygen to make water.

Chemistry is elementary

Today, science has discovered 117 chemical elements, although only 94 occur naturally on Earth. You know many of them: hydrogen, helium, carbon, oxygen, iron, copper, silver, gold, chlorine, sodium and neon are a few of the most familiar.

In chemistry, the atoms that make up these elements are as small as you can get. But they are rarely found sitting around by themselves. They're too sociable for that. For example, almost all the hydrogen on Earth is in water molecules because, as we just discovered, it takes very little persuasion to get banging with oxygen and make up a cosy little threesome.

Atoms are so tiny that it is impossible to see them, even with a microscope. A pinhead is around 4,000,000 atoms wide. (Except that physicists have found that, like clouds, atoms don't really have fixed boundaries, so measuring them is difficult.)

Physicists have also found out how to trump the chemists' work by smashing the atom to make even smaller bits of matter. But let's not get ahead of ourselves. There's lots to find out about atoms before we start breaking them.

Ice breaker

Now we've made a glass of water, what else can we find out about it?

Well, the obvious thing is that, if we make it very cold, our water will become ice. If we then warm it, it will become water again.

In solid ice, our water molecules are pretty much frozen in place, packed tightly together and held by long-lasting bonds they form with their neighbours (although these bonds do allow them to vibrate about a central point).

As they warm up, the extra energy helps our water molecules vibrate more and more until they break free of these stable relationships to form a liquid. The molecules are still packed tightly, but their bonds with their fellows are fickle and fleeting as they jostle about like horny teenagers at a disco. Now they can flow over each other to fill whatever container they're in. (Note that this change of state – from solid to liquid – takes a lot of energy, which is why, by melting, ice cubes can cool a drink much more than a similar amount of ice-cold water.)

Heat your water to boiling point (which takes lots more energy) and the molecules will gain enough oomph to escape the disco and become free spirits – in other words, they form a gas. In this state, our water molecules have broken all their bonds and are really speeding about, so they spread up and out in all directions until gravity hauls them back or they bump into something solid.

Water molecule

Ice   Water   Water vapour

You can tell that gas molecules are much more spread out than those in solids and liquids because gases are relatively easy to compress (it only takes a bike pump). It is almost impossible to compress solids (unless, like sponges, they have a structure that's full of air) and liquids (which is why hydraulic systems work: try and compress oil at one end of a tube – with your car's brake pedal, for example – and it'll push back at the other – making your brake pads grip your brake discs).

Steamed up

The gas formed from water is called water vapour. Even though it took a combined volume of 1,500 litres of hydrogen and oxygen to make our glass of water, when heated that water makes around 500 litres of water vapour. This is because it's not the size of the molecules that determine the volume of a gas, it's purely down to the number of them.

Water vapour is as clear and colourless as air. The steam we see coming from that saucepan (in which we're boiling away our precious cup of water) is slightly different: it, like the clouds we see in the sky, is actually made up of tiny drops of liquid water hanging in the air.

Soon all the water we made will be gone, boiled off and dispersed into the atmosphere. But we can bring some of it back. Water vapour will condense into water on a cold surface, the cold stealing away the energy of molecules that hit that surface, so that they can no longer fly free.

This is good news for us – water is amazing stuff and we're going to need it again in a minute.

States of matter

Solid, liquid and gas are three states (although physicists like to call them 'phases') of matter. But there are quite a few others, too, though they are much less common, and only created when there is enough energy around to break up the atoms themselves. They include plasma, Bose–Einstein condensates and Fermionic condensates.

This knowledge is only useful if you feel like arguing with someone who tells you that there are three states of matter.

Water: strange thing

Almost all solids expand slightly as they get hotter. That's why large concrete structures have those rubbery expansion joints in them. And almost all solids expand a little further as they turn into liquids. But water is a little odd.

Ice does expand slightly as it is warmed, but then it goes the other way and contracts as it turns into water. The shape of water molecules, and the fact that one side of them has a slightly positive electric charge and the other a slightly negative one, means that they can fit together more closely when jostling around in liquid form than as a solid (in which all those positive and negative charges line up in an orderly fashion, like a grid of mini-magnets).

So water expands as it freezes: bursting pipes, but also causing ice to float, so fish can survive beneath it. Water also has many other properties, such as surface tension, which are found in few other liquids, and it is, as far as we know, uniquely suited and vital for life as we know it.

But for our experiment, water is still a good way to find out about solids, liquids and gases. And after all the trouble we had making our glass of water it would be a shame not to use it.

More about molecules

OK, it's not easy to believe that your small still glass of water is full of jostling molecules. But if you mix a teaspoonful of tiny grains of pollen into your water, and then look at them through a microscope the evidence is right there before your eyes. You'll see those pollen grains jiggle as they're battered by the water molecules.

Brownian motion

This is called Brownian motion after Robert Brown, the Scotsman who discovered it: you probably saw it at school. If our water is likened to a horny teenage disco, then the grain of pollen is like a giant balloon on the dance floor that is being continually bumped this way and that, but never very far, by the host of dancers around it.

In fact, molecules are 250,000 times smaller than pollen grains. We know this because Einstein worked out the maths behind Brownian motion, which, if we measure our pollen grains, allows us to calculate the size of the water molecules. You can't see them, but they really are there.

Brownian motion can be seen in gases, too. Smoke particles in air (which is a mixture of 78 per cent nitrogen and 21 per cent oxygen, plus tiny amounts of argon, carbon dioxide and water vapour) move rather like pollen grains in water. And that's just the start of the physical phenomena that can be explained once we realize that molecules and atoms are always moving.

A hot-air balloon fills up and then floats away because of a whole host of these effects. Come on, let's try it. You'll never have a better excuse to fly around in a wicker basket – with just the laws of physics for support and no control over where you go.

A lot of hot air

We've already seen how hot-air balloons rise because they displace a greater mass of air than they contain. Now we can discover why this happens.

Time for a quick thought experiment. And thought only please, because this one is dangerous.

Imagine a sealed box full of air. Inside, free-spirited molecules are shooting about in straight lines at an average speed of 210 m/s (500 miles an hour) until they bang into each other, or, as is more likely because they're well spread out, into the walls of the box. This steady drumming of molecules on the inside of the box is what we know as *pressure*. It's a bit like what we feel from a jet of water, except those free spirited molecules push out in all directions.

Now, if you put that box on the fire, the air molecules will get fired up. They'll move more quickly, so they'll hit the walls of the container harder (free spirits don't like being kept in boxes). The pressure on the walls of the box rises. Keep your box on the fire long enough and bang: it explodes. Told you it was dangerous – you can't lock up a free spirit for ever.

 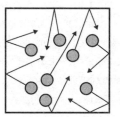

Low pressure   High pressure

Feel the pressure

Pressure is measured in Pascals. It's a measure of the force acting on a given area: so if you halve the area you double the pressure. This is how snowshoes work: by spreading your weight (and remember, weight is force) over a wide area, they reduce the pressure on the snow and prevent you sinking into it. If instead you went walking on the snow in stilettos or on stilts, their tips' tiny surface area would concentrate your weight into lots of pressure on the snow, and you'd fall straight through it.

But we're all under pressure anyway: 100 km of atmosphere presses down on us with 100,000 Pa of pressure. This is why suction cups work: when you press the cup to a surface, some of the air is squeezed out, lowering the pressure inside. Now the higher atmospheric pressure is pushing the suction cup against the surface harder than the cup is pushing away from the surface, so the cup sticks.

A bit about Blaise

The unit of pressure is called the Pascal, after French scientist and philosopher Blaise Pascal. Born in 1623, he did plenty of brilliant maths and science from an early age, including building a mechanical calculator and essentially inventing probability theory. He was also instrumental in the invention of the hydraulic press, so it's appropriate that the unit of pressure carries his name.

After a vision in 1654 and an apparent miracle healing of his daughter in 1657, Pascal gave up all that science stuff and devoted his life to God, while writing arguments that reason was untrustworthy. He died in 1662, aged 39.

Blowing up a balloon

Now we know a bit more about pressure, let's try that box-on-the-fire experiment again, except with some sort of imaginary fireproof party balloon. Make it a red one with yellow lettering that says 'Happy Birthday'.

Once you've blown it up and tied the neck, the pressure inside equals the air pressure outside – that's why it stays at the same size. Now stick this fireproof balloon on the fire for a little while. What happens? As the air inside grows hotter, our free-spirited molecules start hitting the inside of the balloon harder – the pressure rises and so the balloon gets bigger.

Now, as the balloon blows up, the free spirits have more room, so they don't hit the walls quite so often. Eventually – if you take the balloon off the fire so that the air inside reaches a constant temperature – the reduced number of hits on the walls compensates for the extra energy of those hits. Once again, the balloon will reach a constant size. It'll be bigger, but inside and outside pressure will again be equal. All the fired-upness of those free spirits will have bought them a little extra room.

Careful with that fireproof balloon though. That expanded air will be lighter than the rest of the air, and so it will try to float away. Something similar happens to the hot air heated by our hot-air balloon's burner: it expands and rises immediately. This process is called convection and it's a common way for heat to travel. Columns of warm, rising air above hot objects are useful for filling hot-air balloons, or keeping birds of prey gliding while they look for small furry snacks. But they can be a problem, too.

Imagine you're back from a freezing trip in a hot-air balloon. (Did you know that the reason it's cold up high is that the air pressure is lower? Can you see why?) Now you're back inside and there's a nice hot radiator on the other side of the room. So why is there a cold draught on the back of your neck?

Well, the hot air above the radiator rises, and, as it does, it slowly cools. More rising air pushes it away from the radiator and eventually, when it's above you, it has cooled enough to fall – right down the back of your neck. Ugh!

But don't moan. You're not really cold. In a moment, well find out what it means to be so cold that you can't get any colder.

Absolutely freezing

Now we know that temperature is really just a measure of how much the molecules in matter are binging about, we can think about what might happen if they stopped moving completely.

Well, we can think about it, but it can't happen: the third law of thermodynamics says it is impossible for matter to lose absolutely all of its thermal energy and so reach the complete lack of temperature that physicists call 'absolute zero'.

The strange thing about absolute zero is that it's only $-273.15°C$ (and water freezes at $0°C$). That is cold (you'd have to wrap up pretty warm if it was $-200°C$ outside): but when you think that $+273°C$ is a good temperature for cooking thin, crispy pizzas, and the middle of the sun is reckoned to be at $15,000,000°C$, it doesn't seem that cold.

Then again, scientists do have ways of cooling things to temperatures very close to absolute zero, and it is in fact so cold that strange things happen, like, for example, the formation of those Bose–Einstein condensates we mentioned earlier.

Why springs are springy

Before we finish with the hot and heavy stuff and find out what makes light work, we should quickly mention a few of the many other familiar things that are explained by the kinetic theory of matter (which just means that the molecules jiggle around as we've described). Let's look at why puddles evaporate without boiling, why springs are springy, and, most importantly, why some people buy fluffy covers for their toilet seats.

Puddles disappear even on cold days because it's the average energy of the molecules that give the puddle its low temperature. But that average disguises the fact that, at any time, the molecules are actually moving at many different speeds through their packed teenage disco. And the ones that are lucky enough to get enough energy when they're near the surface of the water can – if they get a straight run – fly free, up into the sky, breaking their watery bonds and leaving behind their teenage angst (at least until they turn into rain or condensation). And so the puddle slowly evaporates.

Springs are springy because the atoms in the metals we use to make them are all bonded, like the molecules in ice. Stretch them apart and the bonds pull back. (The molecules in ice are less tightly bonded and the pattern of their bonds is different to that of the atoms in a metal, so if you pull ice it just breaks apart: it doesn't make a good spring.)

And now for a loo break.

Warm and fluffy

The popularity of pink fluffy toilet seat covers is explained by the unbounded strangeness of human taste. But *conduction* and s*pecific heat capacity* are important, too.

Conduction is the process by which the jiggling of one atom causes others near it to get jiggly as well, so that heat travels through matter. This is a familiar process: if you heat a metal saucepan the heat travels through the metal by conduction and heats the water inside. However, plastic is a poor conductor of heat: so the handle remains cool.

A material's specific heat capacity describes how much energy it takes to warm it up a given amount. Water has a high specific heat capacity. This makes it useful in central heating systems, because it means that water can store a lot of heat as it travels from the boiler to your radiator (where it converts most of that heat, via convection*, into a draught on the back of your neck).

OK. Now we know that some materials are better heat conductors than others; and some need more energy to get warm. So let's get to the seat of the problem: that cold plastic toilet seat on your bum in the middle of freezing February night. Yuck!

---

* Although we call them 'radiators', and you can feel the radiated heat from a hot one if you hold your hand near it, most central heating radiators actually transfer 80 per cent of their heat to the room by draughty old convection.

But, quick, try a fluffy pink cover. Mmm. Cosy. And yet both seat and cover are actually at the same temperature: the general, ambient, cold-on-your-bum temperature of everything in your house on a cold night. So why does the cover feel warm? Well, compared to plastic toilet seats, pink fluffy covers don't conduct heat well and do warm up quickly; so they are always lovely and comfortable.

Of course, if you think there's something a bit weird and unhygienic about pink fluffy covers, you could just wait for someone else to warm the seat and dash in as they come out.

A small mistake

In this chapter we've been thinking of matter as made up of little bouncy balls. Well, it's not.

Don't dismiss this idea though. It's what Newton believed. And, as we've just seen, it's a very useful way to think about all kinds of physical phenomena.

It's just that, as we'll see, we have plenty of evidence that, though atoms and molecules sometimes behave like bouncy balls, they are actually made up of lots more tinier bits, and that these bits are so tiny and so different to the matter cluttering up our world that almost everything about them seems strange and wrong.

And to start finding out about the world we can't get our hands on, we need to take a trip to the seaside.

6

Waves

In this chapter, we'll see that all waves are similar: they travel through each other unharmed, they reflect when they bump into some things and bend as they enter others. And yet waves take many different forms: so we'll find out why wave power is tricky to harness, we'll calculate the speed of sound and we'll hold proof that light is a wave in the palms of our hands.

But let's start at the seaside, because it is a great place to find out about the physics of waves. That's because – as you'll have guessed already – the sunshine, the waves on the sea, and the sounds of the seaside are all waves. So come on, grab your surfboard and we'll swim out to discover more.

Once you're through the surf, and sitting waiting for a nice big wave to ride back in on, you'll be noticing the most important feature of water waves: that they lift you up and down.

And there's another important feature of waves that's obvious, too: the waves carry energy to the beach (which is why they can push you along when you find one to ride) but the sea stays where it is. Sure, the surf plops up the sand a bit, but the movement of the waves and the energy they carry (towards the shore) is quite independent of the movement of any bit of the sea (which basically goes up and down with a bit of backwards and forwards so that each point in the sea actually goes round and round in a circle).

So there you go. You haven't even had to show off your surfing skills and you've discovered what a wave is: a transfer of energy without a transfer of matter.

But don't ride in yet. There's lots more we can discover out here about water waves, and, because all kinds of waves do the same wavy tricks, about other kinds of waves, too.

Interfering busybodies

For example, while you're out at sea moving gently up and down, you can make your own little waves by dipping your hand in the water. These ripples spread out on the sea on top of the bigger waves that you aren't riding. So it's clear that lots of waves can travel through the same stuff, in lots of directions, and all at the same time. They don't bump into each other like balls.

But they do interact in another way. Your ripple makes the big waves a little higher in some places, and a little lower in others. Ripples on a big swell are hardly noticeable, but if you drop a couple of surfers into some reasonably calm water it's quite clear that the two sets of ripples make a new pattern as they spread out and collide with each other, because in some places you get two troughs together (and the result is a doubly deep trough) and in others you get two peaks (and the result is a doubly high peak).

In yet other places a peak in one surfer's ripples meets a trough coming from the other surfer and then at that point the waves cancel each other out. It seems as if there is no wave there – but of course the wave energy from both surfers is still travelling through that point.

Physicists call this interaction of two or more sets of waves *interference*. Those two poor surfers we chucked in might have a slightly stronger word for it.

Waves from first surfer

Waves from second surfer

Interference pattern from the two sets of waves

Sea power

Waves that are 1 m high hit every metre-wide section of beach with enough power to run ten one-bar heaters – which is why it is frustrating that we haven't been able to capture wave energy to generate electricity.

Part of the problem with wave power is the way waves carry it. The energy of a wave is proportional to the square of its height. So a 3-m-high wave has 3 x 3 = 9 times more energy than a 1-m wave. And when a hurricane brings 30-m waves, with 900 times the energy of the tiddlers you were generating power from, your clever wave-power gadget is turned into something mangled and useless.

Breaking waves

Waves travel at different speeds in different things. And shallow water is a different thing to deep water. Waves notice this difference, because as the water gets shallower they start scraping along on the sand.

This slows them down, but the slowing is more pronounced at the bottom where the sand is, and has less and less effect the further up the wave you look. So the top of the wave carries on at its same old speed, and so the top starts to move ahead of the rest of the wave below, and eventually it moves so far ahead it falls over, and the wave breaks.

Turning a corner

One result of a wave moving from the deep sea, where it moves at one speed, to a shallow bay, where it moves more slowly, is that so long as it hits the boundary of deep and shallow at an angle, it will turn a corner.

This makes sense if you think of our breaking waves. Just as the top wasn't slowed as much as the bottom and so the waves broke, if part of a wave has reached shallower water that part will slow down, whereas the part that is in deeper water will be moving faster, and so the wave will change direction. This is called *refraction* of the wave.

This situation is a bit like the way in which a bulldozer steers by making the caterpillar tracks on one side move more quickly than those on the other. If its right track turns more quickly than the left, that side of the bulldozer moves forward further in any given amount of time than the left, and so the bulldozer turns to the left.

Mind the gap

We've already found one way to get waves off the straight and narrow, but there are others, too.

On a stormy day, with huge waves rolling in from the ocean, it's clear that the waves make it in through the small gap in the harbour wall and spread out, gently rocking all the boats as they sit at their moorings.

One reason for this is reflection. When wave energy hits a hard wall it has to bounce off; so the wave is reflected and travels back through itself. Throw another surfer into a calm harbour and you'll see the ripples he makes bounce off the walls and spread out.

But if the waves are coming straight through a reasonably narrow gap in a harbour wall you can see another wave effect: it's called *diffraction*. Instead of just the portion of the wave that hits the gap carrying on inside the harbour while the rest of the water remains calm, the wave spreads out a little beyond the wall, and that spread grows the further the wave travels from the gap. The smaller the gap, the greater the effect.

Waves diffracting in a harbour

Transverse and longitudinal waves

Water waves are *transverse* waves. They involve something moving up and down as the wave goes along. If you look at them side on they make that familiar wavy line, a bit like a long train of camels.

Longitudinal waves, on the other hand, involve something moving backwards and forwards in the same direction as the wave travels. But just as with transverse waves, the matter itself doesn't go far; just back and forward a little as the wave travels on.

Although these two kinds of waves might seem quite different, they do both display all the strange wavy properties we've been looking at. Both kinds of wave will reflect, refract, diffract and interfere.

Sound thinking

Sound is a great example of a longitudinal wave. You can picture what's happening if you pick a note on your guitar.

OK? Ding! See, your G-string is vibrating; moving quickly towards you and away again. And, as it comes toward you, it squeezes up the air molecules, creating a 'compression'. Then, as it moves away, it squeezes the air molecules on the other side, leaving the air on your side slightly spread out (and so at slightly lower pressure – a 'rarefaction' or 'expansion').

Then it comes back and you get another area of squeezed-up (or slightly higher-pressure) air travelling your way as the laws of conservation of energy and momentum ensure that the movement of one molecule is transferred to the next.

A tuneful note

Even though you may not be the best guitar player in the world, the regular vibration of that G-string, back and forth 196 times a second, means it makes a nice musical note.

Busy physicists don't have time to say 'times a second' every time they talk about how fast a guitar string is vibrating or a wave is going through its full cycle from peak back to peak, or compression to compression, so instead they say talk about *frequency*, which they measure in *Hertz* (written as *Hz*).

So the open G-string on a guitar has a frequency of 196 Hz; those surfers still waiting out at sea have a frequency of 0.1 Hz (i.e., every 10 seconds they are at the peak of another wave).

Now, if you press down on that G-string halfway along, it will, when plucked, vibrate twice as fast – moving back and forth 196 x 2 = 392 times a second. Musicians will be able to tell that this new note is still G, but a different G, one an octave (that's eight notes of the major scale) higher.

So that's maths and music in perfect harmony. If you halve the length of the string you double the frequency of the sound.

The Doppler effect

We've all heard the way ambulance and police sirens sound different coming and going. And it's because of the *Doppler effect*: a wavy phenomenon named after the Austrian physicist who suggested, long before sirens were even invented, that it would exist.

Think of a typical teenager's car, the sort that's as much a boombox on wheels as it is a mode of transport. And let's imagine that, every second, it throws one pulse of high-powered bass in your direction as you sit there on the beach muttering to yourself about kids these days.

Now, let's say our boombox-car is moving so quickly towards you that every successive pulse is released at a distance so much closer to your ear than the previous one that it will spend half a second less travelling that ever-decreasing distance to you than the previous one.

Doppler effect

Can you hear what's happening? Instead of your ear picking up a pulse every second, it receives one every half a second. That's twice as many pulses per second: which means you hear that constant bass rhythm pumping twice as fast as the driver does (and the note will sound higher pitched). Likewise, after it drives past you'll hear the rhythm slow to half of what the driver is hearing (and the pitch of the sound will be even lower than what's rattling his fillings).

There is one important footnote to the Doppler effect: what we hear depends on the fact that the speed of sound in air is always the same.

Sound waves from something coming quickly towards you don't travel any faster than those from something that isn't moving. They're not like balls thrown forwards from a moving car. This means the speed of sound in air is what's called a *physical constant*.

Measuring the speed of sound

It's easy to measure the speed of sound: all you need is a nice big wall with lots of empty space next to it (try a building next to a field), a stopwatch, a stopwatch-operating assistant, and a sense of rhythm.

Pace out a distance of 100 metres from the wall and stand there. Clap once, and wait for the echo (which is caused by the sound wave reflecting back to you off the wall). Now clap again and set up a nice steady rhythm: *clap, echo, clap, echo.*

Hand your assistant the stopwatch and ask him to

---

time ten complete clap–echo cycles. They'll probably take round about 11.66 seconds.

Now, from clap to you hearing the echo the sound has travelled 2 x 100 = 200 m. And because your sense of rhythm makes you wait that same amount of time again before clapping again, it's as if the sound has gone there and back again.

So, in one clap–echo cycle the sound travels 2 x 200 = 400 m. In 10 cycles, it travels 10 x 400 = 4,000 m. So the speed of sound is 4,000 ÷ 11.66 = 343 m/s.

I hear thunder

So, 343 m/s is quick, but not that quick: it takes 3 seconds for sound to travel a kilometre, whereas light takes only 0.000003 seconds to cover the same distance. That is why we're quite used to a thunderclap coming along a few seconds after we see a lightning flash. And why, in large sports stadiums, you can hear the announcer's voice on the other side lagging behind the version coming from the speakers nearby.

But this is just the speed of sound in air; it's quite slow because air, as you know from the last chapter, is made from lots of spread-out, binging-about molecules. In something where the atoms are packed much more tightly and bonded together, like steel, the movement of one atom affects its neighbours more quickly.

And so sound travels through steel at around 6,000 m/s. That's quick: but nothing like as quick as a certain other wave we've been enjoying on the beach.

Electromagnetic waves

After it was discovered that electricity and magnetism were related (see also chapter 7), in the 1860s a Scottish mathematician called James Clerk Maxwell created and published a set of equations, known today as Maxwell's equations. These equations showed how electric charges create electric fields, and how electric currents create magnetism. Together they showed how electric and magnetic fields travel through space in the form of waves, and how they move at the speed of light (and how fast that speed is).

This was pretty good stuff, but best of all they showed that light – the light we see, that brings us so much information from the world – is purely a combination of electricity and magnetism. That's why physicists call light an *electromagnetic wave*.

To us, it is special because we can see it. But it is only different to many other types of electromagnetic wave in its wavelength and frequency. The radio waves that bring us music and chattering DJs, the microwaves bounced around in microwave ovens to heat food, warming infrared, ultraviolet (that makes phosphor glow), X-rays (which enable us to see the bones beneath our flesh) and dangerous gamma rays are all the same kind of thing. They're all electromagnetic waves like light.

Long-wave radio starts at a frequency of around 10,000 cycles per second (or Hertz) where the waves are a kilometre long; whereas high-energy gamma rays go up beyond 10^{24} Hertz (with a tiny wavelength of 10^{-16} m). Somewhere in the middle are the frequencies that we see as different colours. Red light has a frequency of around 4.4×10^{14} Hz; blue a frequency of around 6.4×10^{14} Hz.

The more you think about this, the more amazing it is. It's worth repeating: all these invisible forms of electromagnetic radiation — which is created by the stars and heaters and radio transmitters and in many other ways — are basically the same kind of thing as *visible light* (which is better known as 'light' because obviously invisible light isn't really light at all). And this invisible radiation is all around us and even, in the case of long wave radio, inside us, passing through us and brick walls as if they weren't there. (Luckily, lethal short wave gamma radiation is quickly stopped by solid objects.)

Enjoying the rays

Light travels at 299,792,458 m/s in a vacuum. That's really quick. It's also a universal constant. It's always that fast, no matter where you are. That's why the length of a metre is based on it.* It's also, as we'll discover in a few chapters' time, one reason Einstein discovered special relativity.

For the moment though, we'll note that all this rushing about makes it much harder to get a handle on light than on sound, which is one reason physicists had a long argument about whether light was in fact made up of waves or particles.

In the dark

The Greeks started the confusion. They thought something came out of our eyes and lit up what we looked at. Explaining why we couldn't see in the dark was always tough for the Greeks.

In 1670, Newton said light was made from particles (or, as he called them, *corpuscles*). That made explaining reflection and the fact that light travels in straight lines and through vacuums easy (especially for Newton, who, as we've seen, was very good at the physics of motion).

* In the eighteenth century, the French defined the metre as one ten-millionth of the distance from the equator to the north pole. Unfortunately, it was later discovered that their metre was actually one-fifth of a millimetre wrong. So we now have to settle for an even less handy definition: one metre is the distance light travels in a vacuum in 1 divided by 299,792,458 seconds. Or we can estimate that a metre is about a yard, which is about one long stride.

Light could now travel from the sun, bounce off things around us and then enter our eyes. In the dark, there was no light, and so nothing reaching our eyes. But Newton did find it tricky to account for refraction and our other weird wavy properties.

Other scientists, including the Frenchman René Descartes (of 'I think therefore I am' fame) and the Dutchman Christian Huygens, believed that, because it could be diffracted and refracted, light was more likely to be made up of waves.

The debate raged for over a century, until a bright Englishman called Thomas Young, who could speak fourteen languages by the time he was fourteen, came up with the not very grandiosely entitled *double-slit* experiment.

Fringe benefits

What Young did was to use light waves of a single colour, and *two* very narrow slits, a tenth of a millimetre apart, for the waves to diffract through, instead of a single gap. (The thinness and closeness of the slits is important because visible light has a wavelength of around 0.0005 mm.)

The result was that, because there were two slits for the light to come through, it was as if Young had created two separate, but identical sources of light. And when the light waves spread out from these slits by diffraction, they interfered just like water waves, creating some areas where the waves reinforced each other, and others where they cancelled each other out.

And, best of all, this interference could easily be seen with the naked eye. In a dark room, Young's slits make a pattern on a screen: a central bright spot, with successive fainter blobs of brightness (called *fringes*) slowly fading out on each side. Physicists can even use the spread of this pattern to calculate the wavelength of light.

So there you have it: one light experiment, two wave-related effects. Particles can't diffract or interfere, so poor old Newton's theory was blown out of the water.

If he hadn't been dead, he'd have been furious.

Spectacular light show

Here on the beach, the fact that light is a wave is obvious.

Let's start with those binoculars that you're using to study the fine array of surfing talent sitting out there on the waves. They use lenses. And lenses make things seem bigger by refracting the light so it bends, because light travels more slowly in glass than in air.

Also, can you see that spray bow above the waves? (It's that rainbow formed in the sea spray when the sun is behind you.) It's there because the white light coming from the sun is really made up of many colours. And light of different colours is refracted different amounts (because it has different wavelengths). So when white light goes into a drop of water (and refracts), reflects off the back of the drop, and then comes out the front (and refracts again), this double refraction causes it to be split so much that a rainbow appears.

The colour of the sky is also a clue to light's waviness. The sky is blue because blue light, which has a shorter wavelength than red and yellow light, is, because of that wavelength, scattered by the air molecules binging about in the sky. So, whichever direction you look (so long as it's up), some of this scattered blue light reaches your eyes, and the sky looks blue. Most of the red and yellow light travels directly to the surface of the Earth and is reflected and absorbed by the sand, the sea and, of course, that surfing talent.

The clincher for light being made of waves is this: you can do your own single-slit version of the double-slit experiment. But we'll get to that in a moment: after a quick look at how the waviness of light could help us all disappear.

Can I be invisible?

In theory, if you bend light around something and then straighten it out again afterwards, that something would become invisible. That something would be a bit like a rock in a stream: a little way downstream of the rock the water has settled down and there is no way to tell (by looking at the stream) that it ever passed a rock.

Theory, of course, is one thing; practice, quite another. But in 2006 and 2009, teams of scientists in the US and the UK made newspaper headlines when they claimed that they really had made objects invisible using light-bending cloaks made of 'metamaterials' – substances engineered to have a precise structure, such as tiny hairs or holes, on a scale similar to the wavelength of light.

However, the first 'cloak' was actually a heavy and

completely-not-invisible metal cylinder that only bent microwaves. The second was almost invisible – but only because it was a hundredth of a millimetre wide. And again, it had no effect on visible light. It only worked on infrared. Many other scientists pointed out that it is difficult to see how these 'cloaks' could ever work for more than a few very specific wavelengths (and therefore colours) of light. The metamaterial would have to be very clever, otherwise, like raindrops, it would split white light into a spectrum as it bent it, and you'd end up hiding under a rainbow.

So, for the foreseeable future, invisibility is out of reach for us mere mortals (though as 90 per cent of the universe is invisible anyway – see page 179 – perhaps that doesn't matter).

Two fingers to light particles

Young used two slits because it makes the interference caused by diffraction clearer. But you still get an interference pattern with a single slit because the light from different sides of the slit travels different distances and still arrives in or out of sync. So here's the handy version of the experiment.

Ready? Hold two fingers very close together so there's a tiny gap between them that you can barely see through. Now look towards something bright, like that white sea wall (not the sun). And there (by the laws of physics and the fact that light is a wave), in the gap between your fingers and running parallel to them, will appear faint dark lines of destructive interference ('dark fringes'). You can see they aren't down to blurred vision if you move your fingers apart a bit further and watch them disappear.

Isn't that amazing? And isn't it even more amazing that Newton's theory of light being made up of particles was believed for a hundred years?

Well, yes, it is. But there is one thing even more amazing than that: in chapter 9 we will discover that Newton was, in fact, and after all the argument, right. Light might behave as a wave, but it *is* made up of particles; only now we give them a phunky name. We don't call them corpuscles (which sound like the sort of nasty blisters that might have afflicted medieval alchemists). We call them *photons*.

But first let's wave goodbye to the physics of waves, an area where pure science is fascinating and illuminating and also very useful.

Uncovering the physics of light has helped us invent and refine telescopes, night vision goggles and the lasers inside CD and DVD players. We've discovered the whole spectrum of electromagnetic waves and seen how light fits into it. As a result, we've built radio telescopes, mobile phones and X-ray machines, as well as finding new ways to kill ourselves (and the bugs that make food go bad) with gamma rays.

The study of sound has helped us design better concert halls, louder loudspeakers and even, if the din is too much, noise-cancelling headphones.

So waves are wonderful. And you'll probably never look at surf in the same way again. All we need now is something to power all our amazing inventions …

7
Electricity

In this chapter we'll learn, amongst other things, about static electricity, batteries, voltage and current, Ohm's Law, the relationship between electricity and magnetism, and DC and AC current. Put them together and you get (a) the building blocks of our powered-up, gadget-mad world and (b) some really dangerous physics.

Today, one form of electricity is regularly rediscovered by shoppers in shopping malls, who get a shock from touching a metal lift button after a couple of hours shuffling around on hardwearing carpets (though it's a mild shock compared to the one they receive when they open their credit card bills a few weeks later).

Electricity of all kinds is essentially the movement of *electric charge.* And electric charge is created by something having more (or fewer) tiny negatively charged particles (called electrons) than it should. (We'll talk more about electrons later.)

Charge builds up because different materials behave in different ways towards electrons. Some, like some carpets, are inclined to collect extra electrons if they get the chance, and so become negatively charged. Others, like leather shoes and human skin, are happy to give away electrons and become positively charged.

The electrical charges that result from this trading of electrons are static electricity – electricity that isn't flowing. But the charge does flow when the unfortunate shopper touches an electrical conductor, such as that metal lift button. This brief flow of charge is called *electrostatic discharge* – the discharge of static electricity. This is what gives our shopper the (small!) electric shock.

Franklin's Kite Experiments

When Benjamin Franklin wasn't plotting with the French to end British rule and found the United States, or working to abolish slavery, he liked nothing better, it is said, than a spot of extremely dangerous practical physics, such as flying a kite into thunderclouds to collect electric charge.

This charge was conducted to the ground along the kite's dampened string, and the tiny sparks that jumped from a key tied at the end showed that this electricity was the same stuff as the static electricity that had been discovered long ago by the Ancient Greeks, who found that their amber jewellery attracted dust because it picked up a charge after rubbing on their clothes. (In fact, the word electricity comes from the Greek work *elektron*, meaning amber.)

Franklin's kite demonstration is so risky we won't even try it as a thought experiment. But Franklin was aware of the danger and he certainly didn't try it when lightning was flashing. In fact, he may never have done the experiment at all, and simply wrote it down as a good idea. Others, including a physicist called Georg Richmann, proved that it wasn't, in actual fact, even a good idea. Richmann was killed by lightning during a similar experiment.

Electric shocks through metal fixtures might be an annoyance to bargain hunters, but understanding why they happen is very useful. Benjamin Franklin's knowledge of electricity led him to invent the lightning conductor: a metal strip that could take the electrical charge from lightning straight into the ground, diverting it before it entered and wrecked tall buildings with fires and explosions.

At the time, churches, as the tallest buildings in town, were often struck by lightning. In fact, at one time people were advised to escape from thunderstorms by going anywhere *except* a church. Today lightning conductors are a common sight on tall buildings, and ships and aircraft have lightning protection that works using the same principle.

The electronic age

It wasn't until 1896 that electrons were discovered, by the British physicist J.J. Thomson. But for a long time it had been known that if you used a powerful battery to give a strong negative charge to a piece of metal at one end of a tube containing nothing but a vacuum (i.e., containing nothing at all), and a strong positive charge to a similar piece of metal at the other end of the tube, rays of something mysterious would shoot from the negative cathode to the positive anode.

Thomson showed that these rays were made up of tiny particles 1,000 times lighter than a hydrogen atom, and that they were the same no matter what substance you made the cathode from. He also spotted that these same particles could be produced by shining the right kind of light on certain metals and by radioactivity. They came to be called electrons.

But plenty of inventors hadn't waited for the physicists to 'discover' what they had been playing with for years. The first electric light was demonstrated in 1803 by the British inventor Sir Humphrey Davy, and by 1880 Thomas Edison started selling practical light bulbs in the United States. Soon electric gadgets were helping us get around, get fed, clothed and watered – and get out the message about just how many new kinds of electric gadgets could be built in the bright new electrical world.

The only problem was that all these new gadgets needed power. Static electricity is good for electrocuting kite fliers and making mall shoppers argue about who touches that lift button, but it is only when electric charge flows along conductors, such as copper wire, that it is useful.

What was needed was a way of getting all those electrons moving – to create electric current.

When life gives you lemons, make batteries

It's quite easy to make a simple battery cell with some ordinary household objects. Take a lemon and stick a piece of zinc in near one end (we'll use a galvanized nail). Stick a piece of copper (like a copper coin) in near the other end. Make sure both pieces go into the flesh of the lemon, but don't touch each other.

The two metal parts are called *electrodes* – they're the parts that electric current will flow from or to. The zinc (nail) electrode is the *anode*, or negative terminal, while the copper electrode is the *cathode*, or positive terminal. In a circuit, electrons will flow from the negative anode to the positive cathode.

Lemon

The lemon battery works by a chemical reaction – the acid in the lemon dissolves the zinc, releasing electrons at the anode and thereby making it negatively charged, while a copper–acid reaction at the cathode absorbs electrons, giving it a positive charge. Unfortunately this battery is very weak (lemon acid isn't very strong, after all). So weak, in fact, that to see it working you'll need to either use an electronic multimeter, or make several lemon batteries and use them all at once to power something. But don't expect them to light up your life – you'll need nearly 10,000 to power a torch bulb.

Ohm time

The laws governing current, voltage and electrical resistance were discovered in 1820 by a German school teacher and physicist called Georg Ohm, after the Italian Count Alessandro Volta helped him out by inventing basic lemon-like batteries.

Ohm found that a battery forces charge to flow through a circuit because it applies a *voltage*. The voltage of a battery is, in fact, the difference in *electric potential* between its electrodes. Electric potential is just a fancy way of saying 'amount of electric charge'. So if you took your lemon battery, and replaced your electrodes with metals that

WHY BALLOONS RISE AND APPLES FALL

produced and absorbed electrons even better than zinc and copper (when you stuck them in a lemon), then your anode would be even more negatively charged and your cathode even more positively charged, so there would be a bigger difference in electric potential, and so the battery would have a higher voltage. Phew! Got that?*

The electron flow around a circuit is called electric *current* – more voltage, all else being equal, means more current. But the electrons don't really want to flow around the circuit – they were quite happy where they were, thank you very much – so a circuit is said to have *resistance*. The more resistance, the less current is allowed to flow.

Ohm's law

Ohm's law is expressed in this equation:

Current (measured in Amps) = Voltage (Volts)/ Resistance (Ohms)

This equation simply states that you can find the current flowing through a circuit by taking the applied voltage and dividing by the circuit's resistance. More voltage means more current, more resistance means less current.

You can think of the whole thing as being a bit like a bike wheel, with the rim and tyre standing in for the electrons.

* Another way to raise the voltage might be to use a stronger acid than lemon juice.

The voltage applied by the battery is like the force applied by the pedals: it causes the whole wheel to turn (so all the electrons in the copper wires begin to move at once, at a speed of a few centimetres per hour). The resistance of the circuit is like your hand slowing the wheel as it spins, with the rotational energy of the wheel being converted into heat by friction so that it warms your hand.

Ohm's discoveries are pretty straightforward. Ohm's law says that if you double the voltage across something that resists current flow (such as the filament of a light bulb) you double the current (which is basically the flow of electrons through the circuit). It holds so long as your circuit doesn't heat up. Then again, if you put current through a light bulb it does heat up, so double the voltage doesn't quite produce double the current.

Also important is the amount of power that the battery delivers to the bulb. This is given by multiplying the current by the voltage. So, if you can double the voltage and the current you can quadruple the power delivered to the light bulb. (Until your bulb burns out anyway.)

Electric Power

The equation for electric power is very simple: Power (measured in Watts) = Voltage (Volts) x Current (Amps).

As with other simple equations, this can be easily rearranged. So Current = Power/Voltage and equally, Voltage = Power/Current.

A good lemon battery might produce a voltage of around one Volt. Join the nail and the coin to the terminals on a bulb and that difference in charge will cause a current to flow. But, as we've seen, a lemon battery can't supply the current (or, therefore, the power) to light a bulb. It can barely make all those electrons in the circuit move.* A single AA-size battery, on the other hand, can get the electrons trundling around the circuit quite happily: it can deliver 1.5 Volts and enough current to light a small bulb for half a day.

Meanwhile, a few hours at the sales in plastic-soled shoes on shopping-mall carpets might give you a static charge of a few thousand Volts as electrons are transferred from the carpet to you. This is enough voltage to create your own tiny shock of lightning – and an aversion to shopping – as you give up the argument and nervously reach for that lift button.

A flow of charge in one direction, as in these examples, is called direct current (or DC). But as physicists studied electricity and magnetism in the nineteenth century, one odd genius, Nikola Tesla, who came from Serbia but settled in New York, would discover that it was much easier to generate and distribute another form of electricity, where the electrons don't flow around a circuit, but instead rock back and forth in one place, creating alternating current (AC).

But first, someone had to discover the link between electricity and magnetism.

---

* Alert readers will be asking where the resistance is to make the current so small. The answer is, it's inside the battery.

The attractions of electricity

In 1820, a Danish physicist called Hans Christian Ørsted noticed that the needle of his compass moved when the electric current in a wire near it was turned on and off.

This was a very important discovery. It showed that, just like magnets, electric currents create magnetic fields. So, for example, the conductors that carry them can be attracted to some metals, such as iron; they can be repelled by other magnets or conductors that have the same polarity (just as the north poles of two permanent magnets repel each other); and they can make pretty patterns with iron filings on a sheet of paper as they move to follow the lines of force in a magnetic field.

The effect is even better if you wrap a coil of current-carrying copper wire around a short iron bar: when the current is on the iron bar is a strong magnet, when it's off, the magnetic field is turned off too. Electromagnets, as these devices are called, are very useful. For example, they make the fabric cones in loudspeakers vibrate (so you can listen to your favourite band and damage your hearing all at the same time), and they pull items containing iron out of household waste so it can be recycled.

The strange thing is that the movement of electric charge actually explains all magnetism. Permanent magnets, like the one in your compass, are magnetic because of the way the electrons in the needle are permanently spinning in sync, a bit like they would be in the presence of an electric current.

Magnetic North is a south pole

The north pole of a magnet is the one that points north; it got its name because that region of the world was already called 'the North'. But actually the north pole of a compass is attracted, like any magnet, to a south pole. So what is called the Earth's magnetic north is actually, magnetically speaking, a south pole. And, due to changes in the Earth's core, it's moving 60 km a year from Canada to Russia.

Don't mention any of this when you're driving somewhere, you're lost, and whoever is navigating isn't sure which way up to hold the map.

Signs of movement

The fact that electromagnets can vibrate speaker cones and pick up iron rubbish gives us a clue that electricity can be used to create movement.

In 1821, a great British experimenter called Michael Faraday was the first to make use of the fact that the magnetic field around a conductor will cause another magnet to turn. But his demonstration of the effect required that the wire carrying the current be dipped in a bath of toxic mercury, which made it an unlikely candidate for the development of electric motors.

Faraday also spotted that the process worked in reverse. If you moved a conductor through a magnetic field, a current would start to flow in the conductor. What's interesting is that the biggest magnet in the world won't make a current flow unless you move it. But the faster you move that conductor, the greater is the current *induced* in the conductor. So spinning a conductor – a coil of wire is best – in a magnetic field is a great way to generate electricity.

All this theory helped lots of inventors improve on Faraday's ideas, until in 1869 Zénobe Gramme, a Belgian electrical engineer who struggled with reading and writing and couldn't do advanced maths, invented the first dynamo that could make enough power to sell. And when Gramme accidentally connected one of his efficient dynamos (that was generating electricity) to another (that wasn't) and spotted that the shaft on the second turned, he realized that he'd also invented the first usefully powerful electrical motor.

Soon, the first commercial power stations were being built. But still, the electric age was not quite born. One man, Thomas Edison, was holding everything up. And part of the reason was that he, too, wasn't great at maths. And part of the reason was that he owned a lot of patents and made a lot of money out of direct current. He didn't want AC rocking his comfortable boat.

Edison tried a lot of stunts, including electrocuting an elephant, to convince people that alternating current was dangerous. He also exploited and ignored Nikola Tesla, who, for a while, ended up digging ditches for Edison's company. But Tesla was a genius. He had done the maths (AC is much more complicated than DC) and his AC designs – particularly of motors, generators and power-transmission systems – won out all over the world, even if no one wanted to build the death rays and another strange machines he proposed in his later years. Edison, who had plenty of other patents and had started movie piracy as a sideline, merely ended up rich (and famous as the inventor of the light bulb – which he didn't invent).

So what was the big advantage of alternating current anyway?

AC/DC: but only AC really rocks

Alternating current is much more difficult to understand than direct current because the voltage is constantly moving between positive and negative, which makes the current flow one way, and then the other. In the UK, the AC that's piped into houses causes the electrons in your wires to rock back and forth 50 times a second. This makes the maths complicated, and causes strange effects. For example, in a thick conductor, AC only flows through the surface of the wire – so the kind of high-current conductors used by power stations are usually hollow.

But alternating current has one huge advantage over direct current. Because it is constantly changing, it is easy to change the voltage with a transformer, and as we'll see, that makes it easier to transmit over long distances, as in, say, a power grid.

Once your power grid includes transformers you can make use of the fact that high-voltage AC can efficiently be sent many miles through the national power grid at 400,000 Volts before being stepped down by a transformer to the 230 Volts that comes into our houses. We use such a high voltage in the power grid so that the electric current in the grid is low.

Remember that electric power is the voltage multiplied by the current. So for a given amount of power, you can

---

Transformers

A transformer is a very simple device, consisting essentially of two coils of wire next to each other. AC current in the first (called the primary) coil creates a magnetic field, which induces an AC current in the second (or secondary) coil. The number of loops in each coil determines how different the current and voltage in the second coil is.

Primary current

Secondary current

Transformer core   Magnetic flux

You can't use a transformer to transform DC current, as it's the constant changing of the AC current that produces the magnetic field.

---

use a low voltage and a lot of current, or a high voltage and not much current, or something in between. But why is it good for the current to be low?

Let's go back to our bike wheel thought experiment. Put the bike in low gear and turn the pedals – this is like using a high voltage to produce a low current. Now place your hand on the tyre to act as a resistance. No problem, right? Now put the bike in high gear – to produce lots of 'current' with a small 'voltage' – and try again. Ow! Friction burn!

It's the same with electric cables. All cables have some resistance, so put a lot of current through them and they'll heat up, throwing a lot of power away. Crank up the voltage, though, and you don't need so much current, so you reduce power losses from heating cables. And the power grid transmits a lot of power, so obviously it will use very high voltages to keep those currents down.

How power stations work

In their most basic form, all power stations use a turbine to convert *kinetic energy* (energy that is due to movement – for example of steam, or wind or water) into electricity. The source of the power – whether it's fossil fuels burned to make the steam, the wind or water flowing through a hydroelectric dam – causes the turbine to move a large magnet around a coil of wire, which, according to the laws of induction, causes an electric current to flow in the wire.

DC power stations, on the other hand, could be located no more than a couple of kilometres from the homes that used their power. Any further away and they'd lose too much power through cable heat loss. Either that or use high voltages that would be very dangerous in the home – at power-grid voltages, electricity will happily jump short distances in air.

So Tesla's AC systems won out, revolutionizing life. From the early twentieth century, you no longer needed one steam engine for every pump and train and machine. The electric age had begun.

8
Relativity

This chapter is all about special and general relativity, the concepts that Albert Einstein used to revolutionize our understanding of everything. Since Newton's time, people had thought the universe was neat and orderly; that it ran like clockwork. But Einstein showed that's just how it seems. If you look carefully, everything is weird – which is why nuclear bombs go bang and you can never beat the speed of light.

Albert Einstein was born in 1879. As a youngster he clashed with the authorities at school. Aged seventeen, he changed nationalities to dodge military service. In 1900, he finally got a degree in maths and physics. But he couldn't get a university job. And so he became a patent clerk.

That hardly sounds like the CV of a genius, but Einstein was stubborn. He knew he had some good ideas and he didn't give up on them. His patent work helped, because he looked at new kinds of electromagnetic devices, but, most importantly, Einstein was very creative.

He knew that he had to think in a different way to solve the problems facing physics. And though the work of other physicists had given him many clues to follow, he couldn't rely on falling apples like Newton. Physics had to go somewhere new. And so Einstein (like us) did physics by thought experiment, looking for general, logical truths about the universe, and trying to understand what they implied.

Then came 1905, Einstein's miracle year. After a day at the office, Einstein would go home to his wife, Mileva, a trained engineer, and to his one-year-old baby son (though he admitted he wasn't a great family man). And there, inside twelve months, he found the energy to write and publish four startling, revolutionary papers on physics.

And the first and most important principle that those papers presented was *special relativity*.

You don't know where you're going

Relativity wasn't entirely Einstein's idea. For example, all of physics is based on the idea that the laws of physics are the same at all times and everywhere. And Galileo broadened this idea so that it covered steady motion.

In the 1600s, the church told Galileo that the Earth couldn't be rotating to the east at hundreds of kilometres an hour because (a) God put man at the centre of the universe and (b) if it was rotating at such a ridiculous speed, any falling apple would move down and to the west. In reply, Galileo told the churchmen to go away and shut themselves in a windowless ship's cabin – and there do a long list of experiments.

As long as the ship moved at constant speed, in any direction, Galileo argued, apples would still fall straight down. In fact, he said, once in the cabin, it is impossible to tell which way the ship is sailing, just as we can't know which way the Earth is rotating, other than by looking at the stars.

Others developed this idea that events are relative to the observers that see them. For example, just because you're in the Back of Beyond doesn't mean you're a long way away from me. I might myself be taking a short break just round the corner, in Beyond.

It's a frame-up

Similarly, if you're driving down the motorway and pass a police car at a speed of 10 m/s (metres per second) you might wonder why it comes after you, sirens blaring, and you get a hefty fine and three points on your licence. But, of course, if the police car was already doing 30 m/s when you passed it, that means, as their clever on-board camera shows, that you were actually doing 40 m/s (and that's almost 90 mph).

Frames of reference

What matters here is the *frame of reference* that the police use to calculate your speed. Sticklers that they are, the police always choose a frame of reference in which the road isn't moving as the basis for their calculation. They are only going to test you for drugs and alcohol if you suggest that nowhere in the Highway Code does it mention the frame reference for the speed limit on a motorway, and that, in an alternative frame of reference, in which their car isn't moving, you were doing just 10 m/s. Or, indeed, that, relative to a frame of reference in which distant stars are fixed, you were actually doing around 30,000 m/s as the Earth moves around the sun.

In physics at least this is a relatively uncontroversial bit of relativity. But that doesn't mean all of relativity has always been uncontroversial. Back in 1905, Einstein was about to tear apart Newtonian physics and make relativity much more interesting.

Lighting the way

As the 1800s ended, it was becoming clear that light was special. It wasn't just that no one could see the aether through which light was meant to travel, and that light's speed (in a vacuum) was always the same: it was also becoming clear that the speed of light itself was important, for example because various equations suggested that it is impossible to travel faster than light. So scientists started looking more closely for evidence that proved aether existed, and to see how it might affect the speed of light.

What's the aether like?

Everyone knows that sound waves travel through air (and other substances) and water waves travel through water – but what do light waves travel through? What is this invisible stuff – which physicists used to call *aether* – that brings the sun's light and energy to us? Towards the end of the nineteenth century, Albert Michelson and Edward Morley spent six years carefully building an intricate device to investigate the aether and answer this question.

They knew that the Earth is speeding through space, spinning and circling the sun. And so it should be travelling through the aether, too. In one direction, therefore, the speed of light should seem faster relative to a scientist on Earth, just as the wind seems to blow harder if you run against it. And in the opposite direction the speed of light should be slower.

But when Michelson and Morley ran their intricate experiment it failed completely. They tried again and again, over a period of years, but with no luck. Whichever direction they chose to look in, the speed of light was exactly the same. There was no evidence at all for the existence of aether.

And so Einstein took a bold step. He ditched the idea of aether and, building on the work of others, made the constant speed of light the centrepiece of special relativity.

First of all, it's worth realizing just how weird this idea is and how it messes up the nice solid Newtonian universe we know and, on balmy days in the apple orchard, generally love.

Let's get you off Galileo's boat and onto a train, because Einstein used trains in his thought experiments. Let's say your train is moving east at 30 m/s. But that doesn't matter to you. Whether you throw your apple forward or back it moves away from you at 10 m/s.

Now, as good old Albert watches from the platform, he sees your apple move at different speeds backwards and forwards. Relative to him, when you throw your apple forwards, it moves forwards at 30 m/s plus 10 m/s, which is a whopping 40 m/s. But when you throw it backwards, it merely moves forward at 10 m/s less than the train's speed, i.e. at 20 m/s.

All this is fairly obvious. But Einstein said that light is different. If our apple, instead of acting like an apple, acts like a photon of light, it doesn't matter which way you throw it: it always passes Albert, who's still there on the platform waiting for his train, at 30 m/s. What's more, relative to you, every photon of light you throw out, also leaves you at 30 m/s.

If you don't understand this don't worry. It makes no sense with apples. Let's try it with those little packages of light known as photons.

Time to wait

Poor old Albert's been waiting around on that platform for a long time. And, as we'll soon see, he has discovered that time isn't quite as dependable as most of us think. If, for example, that train platform is somewhere in outer space near a black hole (see page 149) – or, alternatively, on just about any British train platform if there's snow about – he could be there approximately for ever without seeing a train.

So let's give him something useful to do: another thought experiment. We need two tall glass boxes, each with two mirrors inside, one on the top face and one on the bottom; one for you and one for Albert. And in those boxes, bouncing in sync between these mirrors, you each have your own tiny photon of light.

Now, if your train is stopped at the platform, and because you and Albert have such impossibly brilliant eyesight, you can see your photons bouncing up and down together. So far, so simple. But what happens if we try the same thing while your train is moving at 90 per cent of the speed of light?

On the fast train

Here you come. (Don't get your hopes up Albert. It's not your train.)

As you look at your box, your photon is bouncing up and down as before. No problem. But what does Albert see? He sees your photon move diagonally, in a series of zigzags. It has to, because after it leaves the middle of the bottom mirror it travels up and hits the middle of the top mirror – and that mirror (along with the train and you and the box) has moved forward a bit as it travelled with the train.

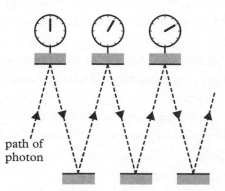

The motion of Albert's photon as he stands on the platform

path of photon

What Albert sees your photon do as you move

But this is where it gets interesting. That diagonal path that Albert sees your photon take is like the long side of a right-angled triangle (because it has gone up and across a little bit). And, of course, this long side is longer than the path you see the photon take, because, for you, it only goes up. In fact, if your train is moving at 90 per cent the speed of light that path is roughly twice as long.

This is much the same as what happened when you were throwing your apples forward at 10 m/s just now. If something moves a greater distance in the same amount of time it must be going more quickly. So if your photon was an apple, Albert would simply see it move faster than you see it move.

Except, of course, light can't go faster. And so something has to give. And, as Einstein discovered, that something that gives is time (and, actually, distance, too, but we'll just worry about time for now and ignore the fact that superfast travel makes you shrink a bit). Time gives up being constant. Because you're moving relative to Albert and because, as Albert sees it, your light has further to travel between bounces (and because it has to travel at the same speed), *time for you must run more slowly than for him.*

Of course, you can't tell that time is running more slowly for you. Your watch still seems to tick as constantly as ever. But, if your train is going past Albert at 90 per cent of the speed of light, as he sees it, for every *five* bounces of the photon in your box, the one in his bounces *ten* times.

And it's not just an effect on photons. Deep in the laws of physics (and Einstein did the maths), time really does slow down as you speed up. So, according to Albert's measurements, your watch would tick more slowly, and if he could see you doing a little dance of physics happiness in time with your bouncing photon, he'd see you doing it all in slow motion.

Even at low speeds this happens. It's just that the effect is so small that we can't see it. But if your train is moving at 30 m/s and you throw your apple forward at 10 m/s, and Albert on the platform measures its speed with his new Super Accurate Apple Speed Measuring Device (which he's just invented), he'll see it moving forward at a tiny bit less than 40 m/s.

So once you accept special relativity, you have to accept that many other things in physics change.

Faster than the speed of light?

Your fast travel slows down your time relative to stationary Albert on the platform – but, relative to you, on the train, it's Albert's time that has slowed. One result of this confusion is that some things, which look like they're happening at the same time to you, will look like they're happening at different times to Albert.

But hold on a minute, you say. This makes no sense and it's all based on not being able to go faster than the speed of light. But who says we can't travel faster than the speed of light? In your thought experiment you've built a super-powerful rocket. And here you are, blasting off into space. What stops you going faster and faster and faster – to the speed of light and beyond?

Well, one way to look at what happens is to look at kinetic energy: the energy a thing has due to its movement. Kinetic energy is related to the mass of an object and its speed squared (see page 54).

According to relativity, the faster you go, the more difficult it becomes to go even faster. But you're still adding kinetic energy with all that rocket power you're using. So if your kinetic energy is increasing and your speed isn't – then your mass must be.

At low speeds, which in relativity means speeds less than a third of the speed of light (which, you'll remember is nearly 300,000,000 m/s), it's almost impossible to notice this effect. But as your rocket's speed increases beyond 100,000,000 m/s (which is much, much faster than any human being has ever gone), all your extra pushing just makes your rocket heavier and heavier and even more difficult to push. And that's why you and I and Albert can never travel faster than the speed of light – even in a thought experiment.

That famous equation

The equation that tells us how energy and mass are related – and how much energy is hidden in all the solid stuff around us – is the most famous equation in physics, which Einstein discovered in 1905:

energy = mass x speed of light squared, or as it is more commonly written:

$E = mc^2$

New killer power

Once you've discover that $E = mc^2$, it is obvious that all you need is a way to convert mass directly into energy and you will get a very large amount of energy (because the speed of light is very large) from a very small amount of mass. Then we can stop drilling for oil and digging for coal and warming the planet and make as much energy as we need, because we can always spare a bit of mass – especially around the waistline.

Unfortunately, controlling that mass-into-energy conversion is tricky, especially as once you find some way of getting the process going, lots of that energy tends to come out as highly dangerous forms of radiation and the result can get rather explosive. This isn't a problem with nuclear bombs, because huge explosions and killer radiation help make these weapons absolutely terrifying. But if you're building a nuclear power station, it's vital that you control the energy and radiation released – and since this is difficult and expensive, nuclear power is never cheap.

The bomb dropped on Nagasaki at the end of the Second World War converted just one gram of mass into energy. But it killed around 40,000 people immediately, and another 40,000 died from injuries and radiation in the following few months. And many more died of cancer and other diseases caused by radiation in the years that followed 1945.

Floating and falling are the same

Making relativity special (and Einstein's other 1905 stuff that we'll find out about in the next chapter) might have been enough for some scientists. But Einstein wasn't satisfied. By 1907 he was making progress on a new, more general kind of relativity theory that included gravity. He called it *general relativity*.

Newton did a good job with gravity. He used the movement of the planets to come up with mathematical descriptions of gravity that are still useful today. And he realized that the force of gravity was a universal. Just as it made the apple fall toward the Earth, it made the moon fall toward the Earth (only the moon keeps on missing because it's going pretty quickly, and so it ends up flying round and round and round). In the same way, the Earth and the other planets in the solar system fall towards the sun, but keep missing.

But what Einstein realized in his thought experiments (and published in his 1916 paper when he had it all sorted out) was that there is no difference between what the moon or a skydiver feels in free fall – and what you feel if we stick you in a spacesuit and send you out of our galaxy into empty space, where gravity is very weak. (Except, of course, that when you're out alone in the space beyond our galaxy you may feel a bit worried about how you're going to get back to Earth given that you can't travel faster than light and, even at light speed, it would take you 30,000 years to get home.)

The similarity of free fall and floating where there is no gravity becomes clear if we ignore the effect of wind resistance and consider what happens to apples dropped by the skydiver and by you in deep space. What happens is that those apples just seem to sit where you let go of them. Alternatively, if you or the skydiver give them a push, they continue on in whichever direction they were pushed.

So, what Einstein realized is that, when you're in free fall and accelerating along with gravity it is impossible to see any effects from gravity. It no longer affects you at all.

And then he came up with another thought experiment.

Sitting down is like blasting off

What, Einstein wondered, is the difference between sitting in a comfy chair in the windowless cabin of Galileo's ship (which is moored in the harbour) and sitting in a similar comfy chair in your similarly windowless spaceship that is accelerating at 9.8 m/s² as you begin that 30,000-light-year journey back home from the Back of Beyond? (You've chosen to accelerate your spaceship at 9.8 m/s² because it's the same rate that the Earth's gravity accelerates falling apples and everything else.)

And the difference that Einstein came up with is that *there is no difference* that you can spot. For example, in both places, if you haul yourself out of that comfy chair and on to a set of bathroom scales, they will say that you weigh the same. And an apple you drop in either place will pick up speed at exactly the same rate: that 9.8 m/s^2.

Einstein explained this with lots of complex maths, but you can get a good idea of his results when you understand that mass, instead of making an invisible force called gravity, instead makes space and time curve. Every object in the world curves space and time, and big things curve it more, and these curves are like pits around the objects that create them. The curvature of time and space – the pits, as we've labelled them – then cause other objects to move (as if they were rocket-powered) because they slide into them.

How space is curved by a massive object

Falling into bed

One way to picture the curvature of time and space and the movement that it causes is to think of space as a cheap old mattress on a cheap old bed, the kind that sags a lot when you lie on it.

Whoever gets into bed with you will enter your gravitational field (which is the bit of space and time, i.e., mattress, you've curved). Once that person is on that curved bit they'll slide into your pit and there'll be two of you down there, and with their mass added that gravitational pit will be even deeper and more difficult to struggle out of than it was.

If a couple of children, the dog and Grandma then also climb on the bed and roll into the pit you've made, then together you've pretty much created your own black hole: an area of space and time so curved by a huge and compact mass that it is impossible for anything – even light – to escape its gravity.

I don't believe it!

And neither did many scientists in the early 1900s. After Einstein, nothing was quite what it seemed any more.

In Einstein's maths, instead of space and time being, well ... space and time, they are just aspects of a single curved, four-dimensional system, which includes our three usual dimensions (up/down, left/right and in/out) along with time. The system sucks up the force of gravity, too. So, instead of the moon orbiting the Earth because of the pull of Earth's gravity, the moon (as far as the moon is concerned) has no forces acting on it and is continuing along a nice straight line. It's just that, with space and time curved by gravity, that nice straight line is basically a squished circle.

And mass, for such weighty stuff, is no longer solid, dependable, or necessarily that weighty. Mass, energy and – it turns out – momentum, are simply three different ways of looking at the same thing. And so all three (and, to make it even more complicated, the related quantities of internal pressure and tension) are sources of gravity.

We've already seen how going fast makes you massive. So does being energetic or under pressure. It's almost as if you go on a diet and turn all your excess weight into energy and higher blood pressure and find that, because energy and pressure have mass, you weigh exactly the same as before you took out that expensive gym membership.

Yes it's complicated and unintuitive, and some people didn't like it, but the predictions that Einstein's theory made, and the way it agrees with all kinds of measurements and experiments, are remarkable. It simply has to be right.

Bending starlight

Newton's law of gravity predicts that the planets orbit the sun in perfect, fixed ellipses. But general relativity predicts that those ellipses themselves will slowly rotate about the sun. This effect had been noticed in the orbit of Mercury before Einstein explained it, and soon after Einstein published his new theory astronomers were able to start confirming that all planets in the solar system had relativistic orbits.

That was only the start of the evidence. Einstein predicted that intense gravitational fields bend light – and when, in 1919, this deflection was measured, Einstein became an international star and part of popular culture. From then on, you didn't have to be Einstein to know that you probably aren't Einstein. Meanwhile, a number of competing theories from other eminent physicists were binned for good.

General relativity also predicted that gravity – like movement in special relativity – affects the passage of time. The more intense the gravitational field around you, the more slowly time will run for you. Again, this isn't saying just that your clock will run more slowly – though it will – it is saying that time for you really is slower than for someone feeling less gravity. Time spent in a high gravity field is a bit like time spent in a pub doing physics on a pool table. You think you've been in there barely half an hour, with hardly enough time to finish your experiment, let alone your pint, but outside it's six hours later, it's cold, it's dark and your dinner is in the dog.

General Relativity and GPS

The strange way that gravity affects time is well proven. For example, compared with the interminable time you experience while stuck in a traffic jam, time literally runs faster (because gravity is weaker) in the orbiting GPS satellites that help your satnav system get its fix. GPS systems rely on very accurate timing, so the clocks in GPS satellites are built to take the physics into account. On Earth, they tick 39 microseconds a day more slowly than accurate Earth clocks. Then, when they're in orbit, they tick at exactly the same rate.

Black holes and a Big Bang

As well as making the specific predictions that we've looked at, general relativity turned out to apply really well to the universe as a whole. Once it was confirmed, it revolutionized cosmology and astrophysics, the sciences that try to explain the entire universe and how it came to be here.

For example, general relativity is an integral part of the theory that says the universe was born in the Big Bang. It also predicted black holes long before we'd built the technology to go looking for them.

A black hole is a place – a collapsed star or a bunch of them, squeezed together – where intense gravity slows time so much that if you were to fall into one we'd see you move more and more slowly as you approached what's called its *event horizon*. But, even if we watched for ever, we'd never see you actually cross it and fall into the hole. Instead, the light from you would slowly fade and change (it would, in fact, be *red-shifted* as if you were moving away very quickly, see page 172).

However, that doesn't mean that a quick dive into a black hole offers you a way to take time out and come back in a few thousand years. Remember, relativity says that you can't see time speed up or slow down for yourself. So you'd cross the event horizon without noticing anything odd, except that you wouldn't be able to turn round and escape. And then ... well, we're not sure exactly what happens then. But with all that gravity, it's likely to be a crushing experience.

Newton versus Einstein

All in all, Einstein's theories of relativity showed how the laws that Newton and many other classical physicists had discovered were merely *just about* right. And in physics, just about right is, in fact, wrong.

Nevertheless, you can't argue with the fact that Newton created modern physics. Newton's system described the behaviour of the entire cosmos, and while others before him had invented grand schemes, Newton's was different. His theories had maths. They made specific predictions that could be confirmed by real experiments. In a way, Newton was lucky: he made such a big impression because he was first. And, in a way, it doesn't matter that we now have better versions of his just-about-right laws. Engineers and some other scientists are quite happy with just-about-right. Newton's laws, and lots of the maths that he worked out, are still used regularly.

Unless you start travelling faster than about a third of the speed of light, or shrink down to the size of an atom, Newton's laws cover just about all the running, jumping, throwing and hitting you can get away with. Where they fall down is if you want to understand what makes the universe tick and where it comes from.

In this regard, after 1905, everything in physics was different (though it took some physicists many years to accept the new ideas, and some never did). Newton's reign as chief physicist genius of all time was over. He couldn't compete with Einstein, a man who almost single-handedly changed how we think of space and time. Especially when, as we'll find out next, Einstein went on to almost single-handedly change how we think about matter and atoms, too.

9

Quantum Physics

In the first few decades of the twentieth century, physicists made huge leaps forward in their understanding of light and atoms. Einstein started everything off, but this time there was also a host of other brilliant men who helped us realize how, if you look closely, the world is unbelievably odd: elements have fingerprints, energy comes in indivisible chunks, light is both a wave and a particle, and some things can be in two places at once.

At the start of the twentieth century, as we've seen, physics was in a mess. The speed of light was one big problem. But there was also a little problem that was just as big: the atom.

Chemists had produced good evidence for the existence of atoms. And at everyday temperatures, as we've seen in chapter 5, Newton's laws could be applied to large numbers of atoms and molecules to predict the behaviour of gases and solids.

But throughout the nineteenth century, physicists had been discovering areas where the combination of Newton's laws and the kinetic theory of matter (which said it was made up of little bits called atoms and molecules) did not work. For example, it predicted that, if your toilet seat was really cold (around absolute zero) its specific heat capacity would be so low that it would warm up instantly as you sat on it. And even those of us who don't own fluffy seat covers know that can't be true.

These problems led many people to believe that atoms were not real. Chief sceptic was an Austrian physicist and philosopher, Ernst Mach. He argued that, because atoms couldn't be seen, they didn't exist. For him they were no more than an occasionally useful idea.

But Mach was wrong. Clever arguments and fancy philosophy are no match for physics and experiments. And they certainly weren't a match for Albert Einstein.

And yet, in what came to be called quantum physics, Einstein outdid even himself. Some of the ideas about atoms and light and matter that he and a few other very clever physicists discovered are so strange that even Einstein could never really get his head round them. He thought his theory went so far against common sense that it was just a stepping stone to something more sensible, and he spent many unsuccessful years trying to find it.

So don't worry too much if what we discover in the next few pages seems a little odd.

You can't be serious

Einstein was a realist. He believed that there must be a single theory that could prove the existence of atoms and accurately predict how groups of them would behave. And, as the greatest genius in physics, he had the brainpower to start creating it.

All Einstein's 1905 papers in the German scientific journal *Annalen der Physik* caused debate and argument. In the second paper, Einstein did the maths that made sense of Brownian motion. In the third, as we saw in the last chapter, he announced his theory of relativity. In the fourth, he discovered that mass and energy were related by the famous equation $E = mc^2$, and paved the way for nuclear bombs.

But it was Einstein's first paper of 1905 that most shook up physics and physicists all over the world. It was not-very-snappily called 'On a Heuristic [i.e. explorative] Viewpoint Concerning the Production and Transformation of Light', a title which highlighted the fact that even Einstein wasn't entirely convinced by his reasoning. And in it, he set forward the revolutionary proposal that light can be two different, contradictory things at once.

Don't open that oven, it's a can of worms

In 1900, a couple of British physicists realized that the accepted laws of physics predicted that, if you looked into a hot oven, you'd immediately be burned to bits by an infinite intensity of ultraviolet light. This was obviously wrong and Einstein's first paper of 1905 started to explain why.

The only problem was that this explanation seemed more impossible than the paradox it was meant to solve. It took nearly twenty years before the majority of physicists began to accept it.

The biggest problem was that Einstein's new theory suggested that light (and all other electromagnetic radiation) is both a wave and a particle. For a realist in particular this was a terrible result: because waves do wavy things (like pass through each other unharmed) and particles do martial arts on pool tables.

It was about as bad as telling physicists, 'I've found a theory that says something can be in two places at once.' No wonder so many of them hated the idea.

Black bodies

A black body is a physics term for an object that completely absorbs all frequencies of electro-magnetic radiation (for example, it absorbs light, and so it looks black). Black bodies only exist in thought experiments (although hot ovens and light bulbs behave rather like them), but thinking about them produces the maths that helps us work out things like the temperatures of stars.

If you heat a black body, it will also emit radiation. And as it gets hotter, that radiation comes out at higher and higher frequencies. So, at 800°C a black body glows red; at 3,000°C yellow; and at 6,000°C it will be white hot.

You can see the same effect with a regular light bulb on a dimmer, because the bulb's filament acts like a black body. At minimum power, the light from the bulb will appear somewhat red. Slowly turn the power up and as the bulb gets hotter, the light will turn orange, then yellow, then eventually white at full power.

Chunks of energy

It makes sense that a hotter object makes a brighter light because, as we know, hotter things have more energy. So a 100-Watt light bulb is brighter than a 40-Watt. But why doesn't the radiation come out of a black body in equal amounts at all frequencies, so that, to our eyes, it always looks white, whether it's dim or not? Why is it that at lower temperatures lower-frequency radiation is more common, so the black body looks red?

The man who came up with the answer – which showed why ovens don't get infinitely hot – was a German physicist called Max Planck. He spotted that the only way to make the maths work was to assume that the vibrational thermal energy of atoms can only be released as light energy in chunks of certain sizes, and that at higher frequencies these chunks would have more energy, because higher-frequency electromagnetic radiation carries more energy around.

Planck's constant

Planck also discovered that the size of these tiny chunks is given by multiplying the frequency of the light by a very small number, which came to be called Planck's constant. (Physicists write this as $E = hf$.) Planck's constant is usually represented by the symbol h, and its value is 0.00000000000000000000000 00000000006626 Joule seconds (more commonly written as 6.626×10^{-34} Joule seconds).

The upshot of Planck's discovery is that if you have a quantity of light energy and keep dividing it in half, eventually you reach a small amount of energy that cannot be divided any further. Planck called this tiny chunk of energy a quantum.

Now, at high frequencies these chunks become quite large, because the frequency of the radiation becomes large. This is even clear in visible light: the deepest violet light we can see has double the frequency of the darkest red; so chunks of violet light must have double the energy of chunks of red light. Once you know this it is clear that, unless our black body (our oven or bulb filament) is very hot, very few of the hot little atoms in it will have enough thermal energy to make chunks of blue light, and so red light is more common. (Remember that all the atoms in your hot black body have different amounts of energy; but it's the average energy of all the atoms that gives the body its temperature.)

So, by inventing the idea of the quantum, Planck came up with a partial explanation for why hot bodies glow red and really hot ones glow white. But it was only after Einstein got hold of the idea that it became part of our picture of how the world works.

Photon torpedoes

If you shine light and ultraviolet on to a piece of metal, the light can knock electrons out of the metal. In a physics lab, these can easily be collected and the number being knocked out measured.

Back in the 1900s, this photoelectric effect was making

physicists rather uncomfortable. The wave theory of light suggested that as you turned up the brightness of the light you should eject more electrons, because brighter light has more energy.

But in their experiments this didn't happen. If you shine red light on to a piece of potassium no electrons are ejected, no matter how bright you make the light. But if you change the light to violet, you can turn it down low and still electrons will be ejected.

Einstein had the genius to see that, if light was made up of tiny quantum particles (which we now call photons), the photoelectric effect is easy to explain. As Planck had shown, it's the frequency, not the intensity, of the light that determines its energy. So Einstein's 1905 paper suggested that low-energy red photons don't have enough energy to knock any electrons out of the potassium, no matter how many of them you throw at it. But even a single ultraviolet photon has enough energy to knock an electron out.

Testing times

This explanation of the photoelectric effect may seem obvious now, but at first, hardly anyone believed Einstein. One American physicist, Robert Millikan, was so annoyed and so wedded to all the evidence that light must be a wave that he spent *ten years* carefully testing the photoelectric effect in order to find something wrong with Einstein's theory. But his careful experiments only ended up confirming Einstein's ideas in every single detail.

For this useful work, Millikan won the Nobel Prize for

physics. Which just shows, in physics, you can get it right even when you're wrong.

Slowly but surely, physicists began to accept Einstein's radical ideas. It helped that other brilliant thinkers used and developed the idea that light comes in packets called photons to explain many familiar effects, including how light can be produced by various substances even when they're cool, or at least not very hot. (A few examples are neon lights, glow sticks and the Northern Lights – where the dark skies above the magnetic north pole glow with spectacular green and red curtains of light.)

Fingerprint physics

All elements, when turned into a gas and heated, give out their own distinct colour. And if, using a prism, this colour is separated out into a spectrum, there will be bright colours in some places and no colour at all in others: a fingerprint that is unique to every element in the universe. This is particularly good news for astronomers, because it means that, simply by looking at the light that the sun and other stars send us, they can learn a great deal about the substances that make them up.

The simplest case is hydrogen (because, it turns out, hydrogen is the simplest element – it has just one electron scooting around just one proton). If you heat hydrogen until it glows and then, using a prism, spread out the light it produces into what, from a white-hot black body, would be a rainbow, all you see are four bright lines of colour: red, bluey-green, blue and violet. The rest of the rainbow is dark.

But if you then let your hydrogen cool, and shine a white light through it, and then through the prism, you get almost the whole rainbow of colours – except with thin dark lines at exactly the places where the bright bits were when the hydrogen was heated.

So hot hydrogen emits red, green, blue and violet light, and cool hydrogen absorbs that light.

When, in 1885, Johann Balmer, a Swiss school teacher, looked at the frequencies of these four bands of light, he realized that they fitted a pretty simple mathematical formula. (He was also able to predict that other frequencies of electromagnetic radiation would be emitted – as ultraviolet and infrared, etc – and years later, physicists with new technology discovered these.)

Because Balmer's equation was so simple it seemed likely that it had something really important to say about what happened inside atoms. The only problem was that nobody knew what that thing might be for another twenty-seven years, until a Danish physicist called Nils Bohr arrived in Manchester to work with Ernest Rutherford, the man whose experiments had given us a picture of the atom as a tiny positive centre surrounded, at a large distance, by negative electrons.

Inside the atom

After J.J. Thomson discovered the electron in 1897, it was thought that atoms were like a 'plum pudding', in which electrons (the plums) were surrounded by a 'pudding' of positive charge, which held them together.

This changed in 1909 when Rutherford, a New Zealander who came to be known as the father of nuclear physics, directed alpha particles at a thin sheet of gold foil and measured how they were deflected. Under the plum pudding model, most of the alpha particles should have been deflected just a little as they made their way through the pudding of gold atoms. What actually happened was that most went straight through with no deflection, but a very small number bounced almost straight back.

This shocking result led to our understanding of the atom as a small, dense, positively charged nucleus orbited by electrons. Rutherford did the maths and found just how little space the nucleus takes in the atom – it's so small that, if you could crush up all the human race, and do away with all the electrons, and overcome the huge repelling force of all those positive nuclei, you could fit us all into a block the size of a single sugar cube (though he didn't put it quite like that).

The nucleus itself is made up of smaller particles – protons (which are positively charged) and neutrons (which have no charge). The number of protons determines what element the atom is, while the number of neutrons affects how stable it is – too many or too few neutrons and the atom may break apart or *decay*, resulting in radioactivity and radiation.

a tiny cluster of protons and neutrons, one thousandth the size of the atom

electrons

Carbon-12 atom, with six protons and six neutrons in the nucleus (not to scale)

Great explanations

But back to Bohr. The Dane had arrived in England in 1911 with little English, but a big dictionary, the complete works of Charles Dickens and the ability to work very hard indeed.

It didn't take him long to revolutionize our understanding of the atom. He knew that electrons couldn't just keep floating around the atomic nucleus by themselves – there needed to be something to stop them radiating away all their energy (as light, ultraviolet, etc) and falling into the middle with a universe-ending bang.

His first brainwave was to combine Einstein and Planck's idea of energy chunks with Balmer's equation that described the emission spectrum of hydrogen. His idea was that electrons can only sit at certain fixed quantum distances from the nucleus. When they get more energy (from heat, for example) they become 'excited' and move further away from the nucleus, say from level one up to level four. (It's a bit like you having more potential energy when you're sitting upstairs.)

But, like you, electrons can only remain excited (or upstairs) for so long. After a while they fall back to a lower energy level closer to the nucleus (falling, say, from level four to level two), and release their extra energy as a chunk of electromagnetic radiation (in this case as a photon of blue-green light).

Because the various energy levels that an electron can sit in are fixed (like the storeys in a block of flats are fixed), so the packets of energy emitted as they move between them are fixed, too – and therefore so are the frequencies and colours. In hydrogen, for example, the fall from level four to level two always gives off that photon of blue-green light. And this creates the colourful elemental fingerprints that physicists and astronomers see in their experiments.

Exciting times

Bohr's idea also explains why gases absorb certain frequencies of light. It's like emission, but in reverse. As the light photons hit the atoms, only those that have the right amount of energy (and therefore the right frequency of light) can be absorbed. So if a hydrogen electron is on level two and a blue-green photon hits it, it has just the right amount of energy to be absorbed, and the electron will jump up to level four. Yellow photons have the wrong energy for hydrogen electrons, and pass straight on through. (It's a bit like a crowd at a football match. Only the blue-green team's goals get the blue-green half of the crowd excited. They don't even bother with polite applause when the yellows do something good.)

Einstein and Bohr, on the other hand, did receive polite applause for their work on quantum theory. And so did the increasing band of geniuses around them. But this was faint praise for the enormous and complicated advances made in physics as quantum theory was rapidly discovered and refined between 1900 and 1928 (especially, when you consider that many of the contributors – including Werner Heisenberg, Erwin Schrödinger and Paul Dirac – are definitely in the top ten physicists of all time).

For example, Lederhosen-wearing, mountain-climbing

German genius Heisenberg discovered that once you start poking around at the quantum level there are real boundaries to what you can find out about photons and electrons. His *uncertainty principle* says that if you find out where exactly a photon is then you can't tell how fast it is going (and vice versa). And this is not because our measuring systems aren't good enough. Instead, measuring one quantity (speed, for example) must affect the other (position, for example) so it can't be known exactly.

Part of the reason that these men didn't receive the worldwide fame they deserve is that the more you learn about quantum physics, the weirder and more difficult to grasp it becomes. And partly the problem is that the experiments that support it aren't easily explained either. But in 1986, all that changed. Laboratory technology was finally good enough to carry out the most important experiment in quantum physics – the one that, almost on its own, makes sense of the whole nonsensical subject.

Quantum physics laid bare

I'm sure you remember our day on the beach: the sun, the surfers, the double-slit experiment, which seemed to prove that light is made up of waves, not particles, because the light waves interfere and cause a diffraction pattern. You even made diffraction fringes in the gap between your fingers.

But what happens if you try the same experiment with single photons of light? If you take a very dim laser beam that sends out those single light chunks, one at a time, and you let them pass through two slits, what happens then? At this, its barest, loneliest point, does light behave like a wave or like a particle?

Well, the brilliant, amazing answer, the one that brings the abstractions of advanced quantum physics into the real world, is that light really is a wave and a particle: both, and at the same time, too. And if, in a temper, you poke at it to try and find out more, well then it just collapses and is one or the other – depending on what you do to it.

OK. Here's the experiment. We set up our dim laser, our narrow slits, and behind them a piece of sensitive photographic film, so that the arrival of each individual photon of light will be recorded. We turn our apparatus on. The first photon is produced. It speeds through. Bang. It hits the film and leaves a mark. A single mark. It is after all a single photon. Aah! It's a particle!

But where is that mark? Well, you can't predict where it will be. But it is unlikely to be in a straight line from the laser, through a slit, and on to the film. Somehow that confounded particle has diffracted and then interfered with itself and so not travelled in a straight line. What's more, if you leave the apparatus running for a few hours, the pattern that appears on the film is the familiar fringes (see pages 111–2), those exact same regions of dark and light that prove diffraction and interference are taking place.

And whether you run the apparatus for hours and let a million photons through one at a time, or turn up the light and blast them all through in a microsecond, the pattern is the same. More photons arrive where the light interferes constructively, fewer where it is out of sync, and cancels itself out.

The answer is inescapable. Each lonely photon travels

through both slits at the same time and then interferes with itself, just like any good wave! Photons can be in two places at once, BUT (and this is, as you can see, a very big and important but) if you start checking for them at the slits you find they actually go through one slit or the other. And then the interference pattern disappears, because, of course, single photons can't interfere with themselves!

In a way, it makes no sense. But it is what happens. Maybe there is no more to say than that.

But what does it mean?

In quantum physics, meaning is hard to come by because things seem so odd to us. What, we wonder, does it mean for everything to be both a wave that goes through both slits and a particle that goes through one? What does it mean when you say that we can never be sure about where a particle is and how fast it is moving?

It doesn't help that quantum physicists themselves aren't in complete agreement about how to explain their discoveries. The Austrian Erwin Schrödinger got so excited by some of his many love affairs that he came up with lots of important quantum maths. But, in an effort to challenge the way we see quantum physics, he also came up with his famous thought experiment, in which a poor cat, shut in a box and at the mercy of a chancy quantum event, is both alive and dead until a physicist opens the lid and forces it to be one or the other. For years, we were expected to believe that a cat could really be both alive and dead. Today, though, Schrödinger's cat is being retired. Many physicists are coming round to the common-sense idea that, although quantum events can defy logic (so our single photon can go both ways through the apparatus at the same time), cats can't. They are dead and leaving their hair in one place or they are alive and leaving it everywhere.

Of course, our problem with quantum physics may be down to the fact that it is only 100 years old and most of us aren't familiar with it yet. Maybe it's like a new pair of shoes that haven't been worn in. After all, it took centuries for most people to accept that the world is round and that it revolves around the sun, despite there being easily available evidence for those theories even before we invented frequent-flyer programmes and space travel. (For example, when a ship comes over the horizon it doesn't just appear as a dot and grow as it would on a flat planet. Instead, you see its masts first, and slowly the rest is revealed.)

Really weird

Scientists certainly have few problems with the effects of quantum physics. They are still adding to the long list of things that it explains, in physics, chemistry and even in biology, where it is shining new light on how our brains work. Paul Dirac used it to predict the existence of anti-matter*

---

* It sounds like sci-fi but anti-matter is real. It's regularly made in physicists' particle accelerators and there's even some appearing inside you, as radioactive carbon-14 decays and makes (among other things) particles called anti-neutrinos. However, anti-matter is hard to hold on to. When an anti-proton meets a proton in the wall of its box, they destroy each other and you're left with nothing but energy.

before it was discovered in the laboratory. And engineers use the advanced equations when they're building microchips and working on new kinds of computer. We even know that basic light and power switches wouldn't work without quantum physics (and an effect called quantum tunnelling).

So quantum physics is useful and well proven. The only problem is, it still raises questions that even the best and most brilliant physicists can't always answer.

10

The Universe

One of the great things about physics is that, in spite of all the wonders we've discovered (and in this little book we have barely scratched the surface), there is still lots we don't know. There's also lots that we sort of may know but aren't quite sure about: stuff that causes debate and argument at physics meetings and on the Internet – and that leads physicists to spend pots of money building bigger and better particle accelerators so that they can make even bigger bangs and smaller bits when they smash things up.

What's more, most of these hotly debated questions are so hotly debated because they touch on important and fascinating theories about the history and origins of the universe and our place in it. And that's two excellent reasons these questions are worth a look.

So, let's finish up by thinking about, amongst other things, the causes of the Big Bang, the possibility of time travel and the make-up of dark matter: the questions that can flummox even a top physicist.

How big is the universe?

Our sun is one among around 300 billion that make up our Milky Way Galaxy, a vast cloud held together by gravity, that includes all those stars, plus more clouds of gas and dust, and, we assume, many hundreds of billions of planets (though, because planets are small and dark compared with suns we've only been able to spot around 500 so far).

On a clear night, well away from the bright lights of the city, you can often see the Milky Way. It's an arch of white light across the night sky, with its centre around the constellation Sagittarius (the one that's supposed to look like a mythical half-man-half-horse centaur, preparing to fire a bow and arrow; or, for those with less imagination, a teapot). The milkiness of the Milky Way is made up of those hundreds of billions of far-away stars, and we see them form that milky band because the Earth is on the edge of things, near the edge of the flattened disc of our galaxy, in a side road off one of its four major spiral arms.

Our galaxy is huge: it is around 100,000 light years across. In comparison, the sun is just eight light minutes away – and you could fit a million Earths in it. The next-nearest star, Proxima Centauri, is four light years away. But the next nearest galaxy, Andromeda – which was discovered in 1925 by the US lawyer-turned-astronomer Edwin Hubble – is 2.5 million light years off. And, as if that isn't far enough away, Hubble also made the surprising discovery that Andromeda – and everything else in the universe – is getting even further away from us all the time.

However, the actual size of the universe is a mystery. We can see 46 billion light years out into space, but, as we shall see, the universe probably extends far beyond that.

Is the edge of the universe a red-light district?

You'll remember from when we looked at waves that, just after a teenager in a boombox on wheels passes you, you hear the pitch of that bass he's pumping out become even lower because of the Doppler effect (see page 105). As relativity predicts, the same happens with the light from stars that are moving quickly away from us: the light waves from them seem to us to have a lower frequency – and therefore appear redder – than they would have done for someone not moving relative to the star. (Physicists say the light from retreating objects is *red-shifted*.)

It turns out that the further out into space we look, the faster the galaxies are moving away from us, right out to the furthest galaxies we can see, which are now that as-far-as-we-can-see 46 billion light years away. The red shift of the light from these most distant galaxies tells us that they're moving away so quickly that the light we now see left them when they were just 36 *million* light years away. (Beyond these galaxies, we reckon there are galaxies moving away from us so fast that the light from them can't ever reach us.)

This all means that everything in the universe is not just getting further away from us: it's getting further away from everything else, too. So the whole universe is expanding. Which means it must once have been smaller.

And if we run time backwards, we come up with a date, 13.7 billion years ago, when the entire universe was very small and very dense. (The Earth, by contrast, is, we reckon, just 4.5 billion years old.) So we need some event to create that small dense universe and turn it into a rapidly expanding one.

That event was christened the *Big Bang* in 1949 by the British astronomer Fred Hoyle. (Hoyle actually used 'Big Bang' as a snappy way to refer to a theory he thought was totally wrong. Unfortunately for Hoyle, it was his alternative theory that was wrong.) But it wasn't until 1967 that proof of the Big Bang was discovered: it's called *the cosmic microwave background* (or CMB for short).

Why do I feel this warm glow?

If it isn't caused by your enjoyment of physics, it may be due to the Big Bang.

As we saw when we looked at matter, something squashed and dense (like the universe as it was being born) must also be very hot. And hot things give off energy as electromagnetic radiation. And hot things as hot and big as the Big Bang give off a great deal of that radiation. In fact, the Big Bang gave off so much energy that, although it's now quite weak, it is still all around us – as the cosmic microwave background, a slight warm glow that fills all of space (if you look using a radio telescope).

We know the CMB is the afterglow of creation because it's special: it's the same in all directions, and its spectrum, with microwaves as its most common component and little visible light, is that of radiation from a black body (see page 155). Nothing but a hot, dense, uniform (i.e., the same throughout) exploding plasma-thing-that-will-be-a-galaxy could have made it.

So, the universe began with a bang. Then it became a rapidly expanding bundle of dense, hot, opaque gas-like plasma, until, after about 380,000 years, it had cooled enough for atoms to form. At that point, it became transparent, so that all the heat radiation flying about could carry on flying about. And that's what our CMB is: radiation from 13.7 billion years ago – a post-Big Bang glow.

The CMB is almost the same in all directions, but there are tiny differences in temperature here and there, which correspond to tiny differences in density. Those differences were vital, because, as the universe expanded and cooled, the slightly denser areas had greater gravity, so they pulled in more gas and became denser still, and slowly the galaxies and the stars within were formed.

In the stars, 10 billion years of nuclear fusion then turned the hydrogen and helium of the early universe into other elements, until they used up the hydrogen. At this point, most of these poor dying stars would have swelled up into what we call red giants (even though they're usually orange). Then, after just a few million years the red giants would have thrown their outer shells off into the galaxy, spreading around all kinds of useful stuff that gathered to build new, more complicated, stars. And when these new stars and their children exploded, the stardust contained heavy elements such as oxygen, carbon, iron and uranium – ideal not only for building further stars, but rocky planets, and even, eventually, us.

Can we escape?

As we look out into the universe – and we can now see evidence for about 125 billion galaxies – it looks the same in all directions. There is no edge and no centre, and, of course, the uniformity of the CMB also suggests there is no centre. So although the term 'Big Bang' is snappy it does give us the wrong idea.

In a way, the Big Bang wasn't an explosion in space, it was the explosion – and the creation – of space.

Today, it seems likely that, wherever you sit in the galaxy, you will see all the universe expanding away from you. Yet this is not just because all the other galaxies are travelling through space. Instead it is because, as Einstein discovered, space-time itself can change shape, and the expansion of the universe is due to the expansion of space itself.

This is why, when you add up lots of expanding space between us and them, some galaxies can be moving away from us at speeds faster than light without breaking relativity. This is also why, although the universe is probably not infinite, it is unlikely, even if you could go faster than light, that you could ever find the edge. Instead, you'd curve round and (eventually) find yourself back where you started.

What caused the Big Bang?

So far, we have direct evidence for what the universe was like after 380,000 years. Is that as good as physics can do?

Actually, with theory and with experiments in particle accelerators that mimic tiny bits of the early universe, we think we can look back much further. A second after the Big Bang, protons and electrons, and various other building blocks of atoms (which we call *elementary particles*), had begun to appear in a ten-billion-degree-Celsius super-hot soup of subatomic bits. Further back still, after a tiny fraction of a second, the forces and laws that hold the universe together (which include gravity and electromagnetism) had appeared. Back then, our soupy universe was hotter still, though it was starting to cool as some of the huge energy from the Big Bang began to be frozen, by the power of $E = mc^2$, into the first raw building blocks of matter.

And before the soup course there was just the cooker: the Big Bang itself, a great flash of energy that may have been caused by a strange kind of field (that's a field like a magnetic field, not one with grass and a cow it) called the *inflaton* field.*

This inflaton field, which basically came out of nothing, very quickly swelled the new universe from a tiny size (where the random effects of quantum mechanics were important) to one where the randomness was so spread out as to be almost invisible – except as those tiny but important variations in the CMB that tell us how the stars were formed.

Sounds mad, doesn't it? But behind this story is a powerful theory, which was invented by the American theoretical physicist Alan Guth in 1979. And the maths of

---

* That's infla-*ton*, like infla-*tion*, which is what the field caused, but spelled wrong so that it matches the spelling of other similar fields, such as the pro-*ton* field.

the story – and the way it predicts exactly the minute variations in the CMB that we can measure – means that it holds up very well. Today, among physicists, inflation theory is pretty well tested and accepted.

Is this the only universe?

Inflation theory says that the portion of the universe that is observable to us today expanded from a size of 10^{-50} metres (that's 0.00 … 49-zeroes-here-please … 001 metres, which is way, way smaller than an atom) in radius at 10^{-35} seconds after the birth of the universe to almost one whole metre in radius at 10^{-34} seconds. (Just think for a moment about how stupidly fast that expansion of space is. It is, for example, much faster than the speed of light. In fact, each time an inflating universe doubles in size, which happens many times, light doesn't even get time to travel from one side of an atom to the other.)

And Guth's theory is even madder than it seems at first. Guth believes that inflation didn't just happen once. Instead, it is happening continuously, with new universes being born like bubbles in their own tiny nicks of inflationary time, then cooling as their bit of the inflaton field relaxes. And as a universe cools it expands more gently, it invents its own laws, and its own ways of doing things at its own gentle pace – and it is then cut off for ever because space outside it keeps on madly inflating, apart from where it forms who-knows-how-many other tiny bubble universes, far, far away.

Guth even believes you could make a new universe in the lab, from a few grams of matter and a special kind of emptiness called a *false vacuum*. Well, maybe you could, but you definitely shouldn't. The huge energy released into our universe would be like thirty of the nuclear bombs dropped on Nagasaki all going off at once. And that's going to hurt, even if you've got your plastic lab glasses and rubber gloves on.

Was it a bug-eyed monster?

The problem with inflation theory is that it is impossible to test. Whichever Big Bang theory you choose (and there are others), they agree that you can't see beyond the Big Bang, or out of our bubble universe. In fact, some would say, it makes little sense to think of what created our universe. The creation is more like an impassable boundary to our universe than an act in it.

But, if you're bored with physics, you could come up with your own story. Maybe our universe was created by some bug-eyed monster who'd discovered the same laws as Alan Guth, a slimy bug-eyed monster who just happened to have some false vacuum handy. But remember, if it was, that alien got blown up a long, long time ago, and, anyway, there is no possible way for us to know anything about its universe, which may have been very different to ours.

Actually, if Guth is right, and the creation of the universe is basically a free lunch, it seems likely that the vast majority of universes are probably too weird to support anything like life as we understand it. That's because there's only a tiny chance that any universe that coalesces out of the hot, dense post-Big-Bang soup will be as cosy for life as ours.

So although our universe can seem vast and inhospitable, we can also see that, if gravity was only slightly less strong, or if there were five or seven or twenty-seven dimensions, or if *dark matter* was less common, then balloons might fall and apples might rise – or, more likely, galaxies and stars could not have formed and there would be no apples, no balloons, and no us to miss them.

But what is dark matter?

Well, for a start, dark matter isn't dark. It is in fact completely transparent. It's just that space is dark, so we can't see it.

When we look out at the universe and add up all the mass we find, there just isn't enough to make enough gravity to help galaxies form and move like they do. In fact, we now reckon that we can see only 5 per cent of the stuff in the universe. The rest is dark.

There must, therefore, be huge clouds of invisible stuff, stuff that light and other forms of electromagnetic radiation passes straight through because it's not made up of protons and neutrons at all, but that supplies the missing gravity to give galaxies the right structure.

It's not just theory. We've seen light from distant galaxies bent by what must be dark matter. But what exactly dark matter is, and how it came to be all over the place, is a mystery. Physicists can't even agree on whether it's cold, or warm or hot. Worse still, much of the dark stuff in the universe is probably not dark matter: instead, it's dark energy, spread throughout space and, according to general relativity, forcing the expansion of space to happen more quickly than it might. We can measure this increase in the expansion of space. But we don't have a clue about what dark energy is. We know it's dark and we know it's energy and er … that's it.

This is one place where physics is still in the dark ages. Our current best model of the universe – the Standard Model – sheds no light on dark matter, dark energy and a few other important things, such as gravity. So although we have plenty of theories about why balloons rise and apples fall, at their most basic level the two theories are separate. Apples fall because of gravity, which is explained by relativity. But balloons rise because of temperature and pressure, and that eventually comes down to quantum physics and the Standard Model.

Why are so many physicists growing a GUT?

Even slim, fit physicists may be working on a Grand Unified Theory (GUT) or a Theory of Everything (TOE): one that will add in what the Standard Model leaves out.

The Standard Model of physics uses the interaction of tiny particles to explain matter, electromagnetism, the strong and weak nuclear forces inside atoms, and how those things affect each other. Remember all those forces we spent so long discussing? Well, there's no such thing – at least on the quantum scale.

First, there are the twelve *fermions* – the particles that make up matter. What happens when these particles meet and interact with light, and how they join together to form atoms, is then explained by a twelve-strong bunch of force-carrying particles called bosons, which fly about at the speed of light, between the fermions, pushing and pulling them around. The bosons include the photon, the particle that carries what we observe as a field of force between things that have electric charge. There's also an elusive thirteenth boson, the Higgs boson, which – the theory says – gives the other elementary particles their mass. (Physicists hope that their £2.8 bn Large Hadron Collider will help them spot one.)

The problem is that, once you try and add in what the Standard Model leaves out, and formulate a GUT, everything gets much more complicated. Some progress has been made, but as yet there is no evidence at all that gravity depends on a new kind of fundamental particle called a graviton – and it seems impossible, other than in the huge particle accelerators that physicists can only dream of – to find one.

One avenue of research depends on something called *supersymmetry* – which involves adding in a gaggle of extra particles that no one has ever seen and which creates a whole new set of problems that are possibly only answered by yet another theory called string theory. But it relies on notions like eleven-dimensional space and a lot of maths that is so clever and so new that even most physicists can't understand it. And there is no way to test string theory because the effects that it predicts take place at very high energies and over very tiny distances.

There are alternatives to GUTs and TOEs and getting tied up in strings. One was created in the 1990s by Mark Hadley, at Warwick University in England. It's lean and mean and uses Einstein's space-time to explain quantum physics. No one can find much wrong with it. But it's pretty much ignored. The reason? Well, it could be argued that a thousand scientists are working on their GUTs simply because a thousand scientists are working on their GUTs – and because the best way to get funding for your GUT work is for a bunch of other scientists to agree that what you're working on is worth the massive grant your GUT needs.

Of course, this doesn't mean that our best GUTs and TOEs are wrong. Maybe we will find ways of checking them indirectly. Or maybe, as the next question shows, our best physics isn't quite as good as we'd like to think.

Can we travel faster than light?

No. Of course not. At least, that's what Einstein argued. And so it seems we are stuck in our solar system, forever forbidden by the laws of physics from building a starship *Enterprise* and seeking out new life and new civilizations.

But actually, in quantum physics there does seem to be something that happens faster than light: communication between what are called *entangled particles*.

Entangled particles are pairs of particles emitted when one particle decays (either naturally, or because a physicist bashes it) into two new particles, so that if one has spin of up, the other has spin of down.* But quantum physics also says those particles have no definite spin at all until that spin is measured. And yet, once you do the experiment, and Alice measures the spin of one particle, Bob will find that the spin of the other is opposite, even if Alice and Bob are a long way from each other.

* Spin is a strange property of tiny quantum particles. It's complicated and doesn't involve anything actually spinning around.

So it seems that, once the spin of one particle is measured, it immediately communicates to its twin what its spin should be. (And when we say immediate here, we mean immediate. Clever experiments have shown that the communication between entangled particles happens at speeds millions of times faster than the speed of light.)

Physicists tend to argue that actually this experiment does not break special relativity, because nothing in it really moves faster than light. They say that twin particles could not, for example, even be used to communicate information, let alone help something as big as us dash to the other side of the universe through a *Star Trek*-style teleporter.

But that doesn't stop entanglement being a real problem for current physics. And it doesn't necessarily stop us boldly going and meeting Vulcans, either. Because we still we have one last, best hope for intergalactic travel: a science-fiction shortcut called a wormhole.

Can we jump through a wormhole?

Wormholes are space-time tunnels that could, in theory, lead directly to far-off galaxies (or far-off times). By travelling through a wormhole, you could go as far as you want, and arrive even before you left. And the good news for budding Kirks and Picards is that the laws of physics, as we understand them, say wormholes aren't pure sci-fi: they could exist. The bad news, though, is that the laws of physics also say that wormholes must be very unstable, and as soon as you try to stuff anything as big as you, or your spaceship, or even an atom, down one, it will collapse.

Which leads us to the biggest physicist flummoxer of them all ...

What is a law of physics?

The idea that there might be laws of nature originally came from thinking about the laws that societies create. The laws of physics were seen as a vast force of cosmic policemen, keeping order among matter and everything else in the universe.

Christianity sacked the police force, because it had one powerful God to do all the law making and keeping, and because the universe seemed so well behaved. Newton thought that the simple, elegant, mathematical laws he discovered were thoughts in the mind of a straight-thinking God, and that they were nice proof of just how orderly and rational He was.

Today, the fact that the fundamental laws are small in number (even if the quantum ones are pretty weird) are just some of the reasons that few physicists still think we need a god to keep things in order. Instead, the laws have come to be seen as fundamental properties that create the entire universe (and us in it), just like a few simple rules mean water molecules can freeze together in an infinite and beautiful variety of snowflakes.

But the importance of maths in physics makes it easy to dodge the question of what the laws of physics are. Lots of physicists like the slogan: 'Shut up and calculate!' They plug the numbers in: the acceleration due to gravity; the number of apples in the basket; the size of the balloon; the temperature of the air inside. They make the prediction:

balloon to rise to 2,500 metres. They wear that extra jumper. And if, when they have their balloon trip, they're warm and toasty, then they like the law.

But the laws of physics aren't only about getting your sums right. The best laws also explain why the sums work. For example, in 1915, the German mathematician and physicist Emmy Noether discovered how the conservation laws – which, as we've seen, are some of the most basic and important in all physics – arise directly from the structure of time and space.

For example, the fact that the laws of physics are the same *at all times* gives us the law of conservation of energy. And the fact that the laws of physics are the same *everywhere* means that momentum must be conserved. And once you know this, and once you understand the shape of space-time, as Einstein did, special and general relativity fall right out. All it takes is knowledge, creativity and maybe a touch of genius. It's amazing what the human mind can do. And it's even more amazing that we've tested and tested these ideas and they seem exactly right. So maybe that gives us some sort of answer to the question about laws. But it doesn't end the debate.

Noether proved her theorem using maths. And theoretical physicists, like Einstein, routinely come up with new ideas by creating new maths, long before these new ideas can be tested in the lab. So, even deeper in physics than laws that explain the laws is the maths that explains those basic laws. And the reason why, in some deep way, maths matches the way the universe works is the deepest mystery of all.

Are there any more questions?

Of course there are. Books full of them. If you ever get the chance to flummox a real physicist you could be talking for hours about the beauty and strangeness of the world.

And it isn't just the big questions about the fundamental laws of nature and the creation of the universe that modern physicists find fascinating and inspiring. Many are getting to grips with practical problems: they're improving the efficiency of nuclear and solar power, building spacecraft and working out how we can better predict the weather; they're creating tiny machines and harder, lighter engineering materials using nanotechnology; and they're still busy inventing new gadgets and improving the ones we already rely on, from computers to hospital scanners.

But whatever they do, physicists are driven by the lure of discovery: the lure of being, for a moment, the first and only person ever to know some new thing, and, at the same time, to know that they alone have found that new thing – and that they have found it purely through the power of their own thought.

Isn't physics fun?

Further Reading

If you're interested in the mysteries of quantum physics, *Introducing Quantum Theory*, by J.P. McEvoy and Oscar Zarate, is great. It's a brilliant introduction to the discoveries and to the amazing people who made them, with drawings that bring everything to life.

If you want to know more about cutting-edge physics in general, and the really big questions about our place in the universe in particular, *The Goldilocks Enigma*, by Paul Davies, has it all. It's detailed, well written and accessible.

Index

A

'absolute zero' 93–4
AC (alternating current) 124,
    127–30
    see also current (electric)
acceleration and deceleration
    39–40, 41
aether 11, 136, 137
air/ wind resistance 29, *29*, 31
Andromeda 171
angular momentum 72–4
anti-matter 168
Archimedes 23, 32–5, *35*
Aristotle 11, 12, 18
astronomy 12–13, 14, 143, 147
    see also the universe
atoms 27, 152–3, 161–4, *162*,
    174
    conduction 95
    heat 70, 156–7
    structure of elements 83–4

B

balancing forces 30–5
Balmer, Johann 160–1
batteries 120–4, *120*

Big Bang 149, 173–4, 175–8
black bodies 155–6, 157, *160*,
    174
black holes 149
Bohr, Nils 161, 163–4
bosons 181
Brown, Robert 89
Brownian motion 88–9, *88*,
    154
buoyancy 32–5

C

calories 44
centripetal force 36–7, *36*
charge (electric) 116–17, 124
chemical elements 83–4,
    159–61, 164
conductors and conduction
    95, 117–18, 119, 125,
    126–7, 128
conservation
    of energy 45–6, 50, 66–7,
        103, 185
    of momentum 66–71, 73
convection 92–3
conversion of energy 45–8,
    54